FEYERABEND

Key Contemporary Thinkers

Published

Jeremy Ahearne, *Michel de Certeau: Interpretation and its Other*
Peter Burke, *The French Historical Revolution: The Annales School 1929–1989*
Colin Davis, *Levinas*
Simon Evnine, *Donald Davidson*
Andrew Gamble, *Hayek: The Iron Cage of Liberty*
Phillip Hansen, *Hannah Arendt: Politics, History and Citizenship*
Christopher Hookway, *Quine: Language, Experience and Reality*
Douglas Kellner, *Jean Baudrillard: From Marxism to Post-Modernism and Beyond*
Chandran Kukathas and Phillip Pettit, *Rawls: A Theory of Justice and its Critics*
Lois McNay, *Foucault: A Critical Introduction*
Philip Manning, *Erving Goffman and Modern Sociology*
Michael Moriarty, *Roland Barthes*
William Outhwaite, *Habermas: A Critical Introduction*
John Preston, *Feyerabend: Philosophy, Science and Society*
Susan Sellers, *Hélène Cixous: Authorship, Autobiography and Love*
Georgia Warnke, *Gadamer: Hermeneutics, Tradition and Reason*
Jonathan Wolff, *Robert Nozick: Property, Justice and the Minimal State*

Forthcoming

Alison Ainley, *Irigaray*
Sara Beardsworth, *Kristeva*
Michael Best, *Galbraith*
Michael Caesar, *Umberto Eco*
Gideon Calder, *Rorty*
James Carey, *Innis and McLuhan*
Eric Dunning, *Norbert Elias*
Jocelyn Dunphy, *Paul Ricoeur*
Judith Feher-Gurewich, *Lacan*
Kate and Edward Fullbrook, *Simone de Beauvoir*
Graeme Gilloch, *Walter Benjamin*
Adrian Hayes, *Talcott Parsons and the Theory of Action*
Ian Holliday, *Michael Oakeshott*
Sean Homer, *Frederic Jameson*
Christina Howells, *Derrida*
Simon Jarvis, *Adorno*
Paul Kelly, *Ronald Dworkin*
Carl Levy, *Antonio Gramsci*
Harold Noonan, *Frege*
David Silverman, *Sacks*
Nick Smith, *Charles Taylor*
Geoff Stokes, *Popper: Politics, Epistemology and Method*
Nicholas Walker, *Heidegger*
James Williams, *Lyotard*
Thomas D'Andrea, *Alasdair MacIntyre*

FEYERABEND:

Philosophy, Science and Society

John Preston

Polity Press

First published in 1997 by Polity Press in association with Blackwell Publishers Ltd.

2 4 6 8 10 9 7 5 3 1

Editorial office:
Polity Press
65 Bridge Street
Cambridge CB2 1UR, UK

Marketing and production:
Blackwell Publishers Ltd
108 Cowley Road
Oxford OX4 1JF, UK

Published in the USA by:
Blackwell Publishers Inc.
Commerce Place
350 Main Street
Malden, MA 02148, USA

ISBN 0-7456-1675-5
ISBN 0-7456-1676-3 (pbk)

A CIP catalogue record for this book is available from the British Library and the Library of Congress.

Typeset in 10½ on 12 pt Palatino
by Ace Filmsetting Ltd, Frome, Somerset
Printed in Great Britain by T J International, Padstow, Cornwall

This book is printed on acid-free paper.

For my parents

Contents

Preface

Paul Feyerabend, whose productive career lasted forty-five years, wrote on a plethora of philosophical issues. He was no narrowly-trained or narrowly-focused academic. His time as a student was divided between the study of singing, stage-management, theatre, Italian, harmony, piano, physics, mathematics, astronomy, history and sociology. He became a student of philosophy only after having received his doctorate in the subject. His career as a philosopher was an accident, and he did not see himself as a philosopher. Rather like his one-time mentor Karl Popper, Feyerabend seemed to hold the interest and opinion of those who are not professional philosophers in higher regard than those of his academic peers.

This book focuses primarily on Feyerabend's philosophy of knowledge. His work in this area is exciting and important not just because it constitutes a running critique of other philosophical approaches to science, but also because of its aim. Feyerabend's philosophy of knowledge suggests the possibility of freeing our intellectual lives from irrelevant constraints. It urges us to use our cognitive resources to the full, to realize the human potential that once drove the 'scientific revolution'. In this respect, Feyerabend's work both drew from and contributed to the heady climate of conceptual radicalism and social permissivism which bathed the 1960s.

Because Feyerabend wrote so much, and because this book is conceived as a critical introduction to his thought, I have concentrated almost entirely on his own writings, rather than the secondary literature, flagging the latter in footnotes where appropriate. Although Feyerabend's philosophical work is founded upon the extensive study of quantum theory which he made in the early 1950s while he was one of Popper's students, I deal with this material only peripherally because it is impossible to do so in any more depth in an introductory book.

I would very much like to thank the following for their help in various respects: Andrew Wright and Jim Duthie of the University of North London, for encouraging my interest in Feyerabend; Bill Newton-Smith of Balliol College, and Kathy Wilkes of St Hilda's College, Oxford University, for much inspiration, argument, criticism and help; Professor Paul Churchland of the University of California at San Diego, for answering my questions on the contextual theory of meaning and on his own intellectual relationship to Feyerabend; and my colleagues Hanjo Glock and David Oderberg, for reading parts of later drafts of this book. Finally, I would like to thank my department, the University of Reading, and the British Academy for giving me the time to complete work on this material, Martin, for his friendship, and Debbie, for her companionship and encouragement.

Professor Feyerabend was very helpful in giving me references to several of his articles. I like to think that his later self might have enjoyed, and perhaps even endorsed, my critical evaluation of his earlier self in this book.

John Preston
Reading

Acknowledgements

I am grateful to Grazia Borrini Feyerabend for permission to quote from Paul Feyerabend's work; to Cambridge University Press for permission to quote from his *Philosophical Papers*; to Verso (New Left Books) for permission to quote from *Against Method* (1st and 3rd editions), and *Science in a Free Society*; and to the University of Pittsburgh Press for permission to use excerpts from the essay 'Problems of Empiricism', published in *Beyond the Edge of Certainty: Essays in Contemporary Science and Philosophy*, ed. Robert G. Colodny (© University of Pittsburgh Press, 1965).

Note on References

Works cited in the text are listed in the Bibliography (pp. 223–30). Feyerabend's books are referred to by the initials listed there, and his articles are referred to by their original date of publication. Where the reference is to an article reprinted in one of the volumes of his *Philosophical Papers*, the page numbers for the relevant volume (*PP1* or *PP2*) are given instead of those for the original article. Where the reference is to a passage that appears in an original article but *not* in the version reprinted in the *Philosophical Papers*, page references to the original article are given. '*AM¹*', '*AM²*' and '*AM³*' refer to the first, second and third editions of *Against Method*. Unless another author is specified, references of the form '[19**]' are to articles by Feyerabend.

Introduction:
Feyerabend's Life and Work

Paul Karl Feyerabend was born in Vienna in 1924. As a young man he attended high school there, got caught up in events that he did not understand, was inducted into the *Arbeitsdienst* (the 'work service' introduced for all citizens by the Nazis), and served in the German army's Pioneer Corps during the second world war. In 1943, he learned of his mother's suicide. The following year he received the Iron Cross for leading his men into a village under enemy fire, and occupying it. After being shot during the retreat from the Russian front, he was left temporarily paralysed at the end of the war.

In 1946 he received a state fellowship to study singing and stage-management for a year at a *Musikhochschule* in Weimar. There he studied theatre, and took classes in Italian, harmony, piano, singing and enunciation. (Singing remained one of his life's major interests.) Soon becoming restless, he returned, still on crutches, to his parents' apartment house in Vienna. Although he had planned to study physics, maths and astronomy, he chose instead to study history and sociology at the University of Vienna's Institut für Österreichische Geschichtsforschung, thinking that history, unlike physics, was concerned with real life. But he became dissatisfied with the study of history, and returned to theoretical physics. Together with a group of science students who considered themselves superior to students of other subjects, Feyerabend infiltrated philosophy lectures and seminars. Although this was not his first contact with philosophy, which he first encountered by accident at high school, it seems to have been the period which cemented his interest. He later recalled that in all interventions he took the radical 'positivist' line that science is the basis of all knowledge; that it is empirical; and that non-empirical enterprises are either formal logic (which includes mathematics) or nonsense. This is the view associated with the Logical Positivists, a group of philosophers and scientists comprising the 'Vienna Circle', which flourished in Austria from the early 1920s.

In 1948, at the first meeting of the international summer seminar of the Austrian College Society in Alpbach which Feyerabend attended, he met the philosopher Karl Popper, who was to be the largest single influence (first positive, then negative) on his work. Of this he later said:

> I admired [Popper's] freedom of manners, his cheek, his disrespectful attitude towards the German philosophers who gave the proceedings weight in more senses than one, his sense of humour . . . [and] his ability to restate ponderous problems in simple and journalistic language. Here was a free mind, joyfully putting forth his ideas, unconcerned about the reaction of the 'professionals'. (*SFS*, p. 115)

Feyerabend attended the Alpbach symposium many times, first as a student, then as a lecturer, as a seminar chair and, in what he deemed the most decisive step of his life, as the society's 'scientific secretary'. He later traced the origin of his career and of his reputation back to the decision to accept this post.

In 1949 Feyerabend set up the 'Kraft Circle', a university philosophy club centred around Viktor Kraft, the last member of the Vienna Circle remaining in Vienna, and Feyerabend's dissertation supervisor. The Kraft Circle's main concern was the question of the existence of the external world, a question which positivists had traditionally rejected as being 'metaphysical'. Modelled on the Vienna Circle, the Kraft Circle 'set itself the task of considering philosophical problems in a non-metaphysical manner and with special reference to the findings of the sciences' ([1966b], pp. 3–4). In 1949, Feyerabend persuaded Ludwig Wittgenstein to speak to them. He later commented that:

> [W]e thought, in accordance with Kraft's ideas, that it was *possible* to interpret science in a positivistic manner and that such an interpretation did not require an external world . . . Not even a brief and quite interesting visit by Wittgenstein himself . . . could advance our discussion. Wittgenstein was very impressive in his way of presenting concrete cases, such as amoebas under a microscope . . ., but when he left we still did not know whether or not there was an external world, or, if there was one, what the arguments were in favour of it. ([1966], p. 4)

Among the Circle's other visiting speakers were the Marxist intellectual Walter Hollitscher, and philosophers such as Bela Juhos, Georg Henrik von Wright and Elizabeth Anscombe. Feyerabend met Anscombe in 1952, at another Alpbach meeting. Anscombe, who was in Vienna to improve her German for the translation of Wittgenstein's work, gave Feyerabend manuscripts of Wittgenstein's work, and discussed them with him. Feyerabend tells us that he himself *rewrote* Wittgenstein's meandering *Philosophical Investigations* 'so that it looked more like a treatise with a continuous argument' (*SFS*, p. 116)! Anscombe subsequently translated part of Feyerabend's treatise, which was published as his review of the book, one of his very first publications.

At the University of Vienna, although he had originally planned to submit a thesis on physics, Feyerabend swapped to philosophy when he got nowhere with the electrodynamics problem he was calculating. He presented almost all of his draft material to meetings of the Kraft Circle, and the resulting thesis (on the topic of 'basic sentences', sentences which simply record sensory impressions), for which he received his doctorate in 1951, was derived from notes taken at these meetings.

Feyerabend had planned to study with Wittgenstein in Cambridge, and applied for a scholarship to do so, but Wittgenstein died before Feyerabend arrived in England. So instead he studied the philosophy of quantum mechanics under Popper at the London School of Economics between 1952 and 1953. Having been convinced by Popper's and Pierre Duhem's critiques of inductivism (the positivist view that science proceeds via generalization from facts recorded in basic sentences), Feyerabend came to consider Popper's view, falsificationism, a real option and, he later said, 'fell for it' (*KT*, p. 89). Along with his new colleague Joseph Agassi, he applied falsificationism, the view that good science is distinguished by the theorist's production of and determination to test highly testable theories, in his papers and lectures. At the end of his stay, Popper applied to extend Feyerabend's scholarship, but failed. He therefore invited Feyerabend to become his assistant. Feyerabend declined, feeling uncomfortable with the arrangement, and Agassi took the post instead. Feyerabend, meanwhile, returned to Vienna in the summer of 1953 as an assistant to Arthur Pap, who was trying to reinvigorate the doctrines of the Vienna Circle. There he met Herbert Feigl, who was a visitor at the University, and who convinced him that the positivism of Kraft and Pap had not solved the traditional problems of philosophy, and that 'there was still room for fundamental discussion – for speculation (dreaded word); there was still a possibility of overthrowing highly formalised systems with the help of a little common sense!' ([1966b], p. 5).

During this period Feyerabend was very busy, translating Popper's *The Open Society and its Enemies* into German, writing encyclopaedia articles, and conducting a survey of postwar academic life in Austria for the Library of Congress in Washington, DC. But he also felt that he did not know what to do in the long run, so he applied for jobs in various universities.

In 1955, with the help of Popper and the physicist Erwin Schrödinger, he became a lecturer in philosophy at the University of Bristol, an appointment which lasted until 1958. From 1955 onwards he published many articles, mostly on philosophy of quantum mechanics and general philosophy of science. The early ones strongly reflected the influence of Popper, Kraft and Wittgenstein. Feyerabend attempted to combine falsificationism with the 'contextual' theory of meaning which he read into Wittgenstein's *Philosophical Investigations*.

Feyerabend emigrated to the USA in 1959, becoming Associate Professor of Philosophy at the University of California, Berkeley, and then a full Professor in 1962. No longer under the direct influence of Popper, he began to extend, and eventually to slough off, the falsificationist philosophy. A gradual but fundamental reorientation in his attitude towards philosophy of science saw him align himself increasingly with the outwardly historical approach of T. S. Kuhn, and against what he came to think of as 'rationalism', the tendency to find within or impose upon all worthwhile scientific activity a single 'scientific rationality'. Undoubtedly, student radicalism and the Free Speech movement, both centred on the Berkeley campus, were among his other influences at the time. Of this period he recalled:

In the years 1964ff. Mexicans, Blacks, Indians entered the university as a result of new educational policies. There they sat, partly curious, partly disdainful, partly simply confused, hoping to get an 'education'. What an opportunity for a prophet in search of a following! What an opportunity, my rationalist friends told me, to contribute to the spreading of reason and the improvement of mankind! What a marvellous opportunity for a new wave of enlightenment! I felt very differently. For it dawned on me that the intricate arguments and the wonderful stories I had so far told to my more or less sophisticated audience might be just dreams, reflections of the conceit of a small group who had succeeded in enslaving everyone else with their ideas. Who was I to tell these people what and how to think? (*SFS*, p. 118)

Feyerabend took his academic duties and responsibilities increasingly less seriously, and came into conflict with colleagues and university authorities as a result. But his reputation as a provocative and combative philosopher of science was such that even this did not prevent him from holding several appointments, some of them simultaneous. He lectured at University College London (1966–9); the Free University of Berlin (1968–70); Yale University (1969–70); the University of Auckland (1972 and 1974); the University of Sussex at Brighton (1974–5); the University of Kassel, and the Federal Institute of Technology, Zurich (1970–90). During his time at University College London, he met and befriended another major influence on his work, Imre Lakatos. Lakatos genuinely admired Popper and sought to liberalize the falsificationist philosophy of science. Feyerabend and Lakatos corresponded at length, until the latter's death in 1974, which depressed Feyerabend greatly. Only Feyerabend's part of the joint project they were working on, *For and Against Method*, was published at the time.

The liberalization that Lakatos had in mind was pushed to its extreme by Feyerabend. Like many of his contemporaries, he came to embrace the relativist views that there is no single rationality, no unique way of attaining knowledge, and no single body of truth to be thereby attained. He became intensely sceptical about the ambitions and achievements of 'Western rationalism', suspecting it to be the willing tool of Western imperialism. His sympathies came to lie firmly with people marginalized by this intellectual tradition, and he sought to show that many of its greatest intellectual heroes did not play by the standards which the tradition's self-appointed ambassadors advertise. He also sought to downgrade the importance of empirical arguments by suggesting that aesthetic criteria, personal whims and social factors have a far more decisive role in the history of science than rationalist or empiricist historiography would indicate. His 1975 and 1978 books *Against Method* and *Science in a Free Society* famously gave expression to these anti-rationalist themes, and garnered an audience far wider than books in philosophy of science usually have. Feyerabend saw himself as having undermined the arguments for science's privileged position within contemporary culture, and much of his later work was a critique of the position of science within Western societies. But this is not to be confused with a critique of science itself: in his later work Feyerabend usually took the side of scientists, whom he considered to be opportunists, against the prescriptions of 'rationalist' philosophers. He came to be seen as a leading cultural relativist, not just because he stressed that some theories are incommensurable, but also because he defended relativism in politics as well as in epistemology. His denunciations of aggressive Western imperialism, his critique of science itself, his conclusion that 'objectively' there may be nothing to choose between the claims of science and those of astrology, voodoo and alternative medicine, as well as his concern for environmental issues, ensured that he was a hero of the anti-technological counter-culture. He continued producing philosophical papers right up until his death, at his home in Switzerland, on 11 February 1994.

Feyerabend is nowadays thought of as one of a new school of historical philosophers of science which flourished after the demise of Logical Positivism. The usual story is as follows. In the period between the two world wars, the Vienna Circle, together with Karl Popper (their 'official opposition'), sought a 'logical analysis' of science, an analysis which would make sense of the intellectual products of scientific activity in terms of the formal logic recently reinvigorated by Gottlob Frege's *Basic Laws of Arithmetic* (1893–1903) and by Whitehead and Russell's *Principia Mathematica* (1910–13). The (less well-defined) group of 'logical empiricists' who continued this work after the second world war relaxed the discredited dogmas of positivism, but they too used logic in an attempt to give a 'rational reconstruction' of science. These groups dominated Anglo-American philosophy of science until the late 1950s. During the same period, members of a new school of historians (founded somewhat earlier by Duhem, and developed by Alexandre Koyré, Sir Herbert Butterfield, Vasco Ronchi, Anneliese Maier and others) renewed the *historical* study of science. But their work had little impact on the dominant stream in philosophy of science. Not until the late 1950s, when there emerged a new breed of philosopher of science influenced by the later work of Wittgenstein, did philosophy go to history for a more accurate picture of science. The new breed included N. R. Hanson, Stephen Toulmin, Michael Polanyi, Thomas Kuhn, and Feyerabend.

In the case of Feyerabend, this is at least partly a myth. Although he began his career with an extensive case-study of the quantum theory, this was conducted within a set of assumptions about good scientific methodology derived largely from the work of Popper. In this phase of his work it can be said that, like Popper, his disagreement with the Logical Positivists and logical empiricists is not over whether the methodologies they 'propose' for science should be answerable to the realities of scientific practice (all are agreed that they need not), but only over the methodology proposed. Feyerabend acquired the reputation of being one of the new breed of philosopher of science because some observations he made in early papers were originally drawn from or inspired by historians who really *were* members of the new school. In fact, he should have been regarded, at that time, as something of a fifth-columnist. The reality is that only gradually, partially, and rather late in his active career, did Feyerabend become a genuine 'historical' philosopher of science.

Feyerabend now has a curiously double-sided reputation. Within philosophy of science, the material he wrote during his earlier phase is the better-known. During the 1960s he was perceived to be at the forefront of a new wave in philosophy of science, and his writings received a great deal of critical attention. But philosophers of science turned off when, in the mid-1970s, he produced *Against Method*, whose main theme was the non-existence of scientific method, and whose final chapter suggested both that science was continuous with myth, and that it ought to be separated from the modern democratic state in the way that religion already had been: Feyerabend was then perceived to have isolated himself, by his views (and his behaviour, about which gossip abounded) from the community of philosophers of science.

From that point onwards, Feyerabend's work had a rather different audience. Within philosophy it was studied mainly by those interested in pursuing, or destroying, relativism. But, more importantly, his name also became known to people in all sorts of walks of life because of his critique of the claims of science, his defence of cultural relativism, and the support that these views lent to the anti-technological counter-culture that emerged from the 1960s. Critical study of this later work is still in its infancy, mainly because few philosophers take its premises seriously.

This book aims to give a critical introduction to the central themes in Feyerabend's philosophy, in chronological order.[1] His work can be (roughly) divided into two phases, the first stretching from the early 1950s until about 1970, the second from 1970 onwards. Feyerabend often accused critics of finding weaknesses in his work by juxtaposing views he held at different times, and he glamorized what he thought of as his own tendency to move swiftly from one view to another. Because of this, I have sought within each chapter to evaluate views as a package only if they appear in publications produced at about the same time.

The first eight chapters in this book lay the groundwork of Feyerabend's earlier epistemology of science, picking out the themes in his 'model for the acquisition of knowledge'. I try to show how Feyerabend's version of scientific realism, his theory of meaning, his argument for theoretical pluralism and his radical materialist philosophy of mind hang together within this model, underpinned by his conception of methodology. His later work can only be fully understood on the basis of an acquaintance with these earlier articles.

The onset of the second phase in Feyerabend's work is marked by his losing interest in the ambitions of this earlier project. In chapters 9 and 10 of this book the presumption that he defended a unified (albeit developing) model for the acquisition of knowledge is dropped. Whereas the previous chapters cover the central themes in the first phase of his philosophy, in these later chapters I have had to be more selective. I try to set out the basis of his political philosophy, and to explain why relativism, which appeals so much to Feyerabend's later audience, is regarded by many philosophers as untenable.

1
Philosophy and the Aim of Science

1.1 Scientific and Analytical Philosophy

It has not gone unnoticed that philosophy has a deeply self-reflexive quality which sets it apart from other activities, for the nature of philosophy itself has often been an important philosophical issue. In this century the self-conception of Anglo-American philosophy has been shaped mainly by the notions of scientific philosophy, and analytical philosophy.

Paul Feyerabend set out one conception of the subject in one of his earliest papers, where he argued as follows. Philosophy cannot be both scientific and analytical at the same time. If a discipline is to be scientific it must have a certain subject matter, and it must be progressive, in that it will involve coming to know more about the objects which comprise this subject matter. But if we assume that philosophy is scientific in this sense *and* that it consists of analyses (of language, for example), none of its propositions could express discoveries. This is because of the 'paradox of analysis',[1] that any correct philosophical analysis of a concept must be uninformative, and any informative one must be incorrect. Feyerabend concluded that '*philosophy cannot be analytic and scientific*, i.e. interesting, progressive, about a certain subject matter, informative *at the same time*' ([1956a], p. 95). Philosophers must choose between analytical and scientific philosophy.

This conclusion he compared with Wittgenstein's dictum that 'If one tried to advance *theses* in philosophy, it would never be possible to debate them, because everyone would agree to them' (Wittgenstein [1953], § 128), and he criticized philosophers who want a purely descriptive philosophy which nevertheless leads to discoveries and extends our knowledge. But Feyerabend found the Wittgensteinian ideal of pure philosophical analysis unsatisfactory because he thought it unfruitful. So he decided to pursue 'scientific' philosophy, philosophy which, in yielding scientific knowledge, makes progress.

Such a conception, on which philosophy is distinguished from empirical science only by its greater generality, is problematic. The ideal of philosophy not just as contributing to scientific inquiry, but *as* scientific inquiry, is barely intelligible when applied to parts of philosophy other than epistemology (the theory of knowledge) and metaphysics (the theory of what really exists). It threatens to make nonsense of the scope and the history of the subject, and may incur the wrath of scientists, who will accuse philosophers of overstepping their proper domain.

Although at odds with analytical philosophy, this early Feyerabendian conception does not coincide with recent conceptions of scientific philosophy either. Many contemporary philosophers see philosophy as continuous with science, but they do so on the basis of their rejection of any distinction between the *a priori* and the empirical or between the analytic and the synthetic. The analytic/synthetic distinction purports to divide propositions true solely in virtue of their meaning from propositions whose truth or falsity depends on how things are. Analytic or 'conceptual' propositions, the territory of the analytical philosopher, are factually uninformative; synthetic propositions, with which the empirical scientist is concerned, are not. Philosophers who reject such distinctions thereby repudiate the idea of a choice between scientific and analytical philosophy. But Feyerabend, at this early stage, accepts some such distinction in order to choose scientific as opposed to analytical philosophy. One cannot opt for scientific philosophy (in this sense) and yet deny the existence of analytical philosophy, or of conceptual truths, even though one might ignore them for being uninformative.

Feyerabend must, therefore, have had an early change of mind about the analytic/synthetic distinction, for he never subsequently deployed it, and his (very few) relevant published comments consistently deny its existence, as well as the existence of any class of non-empirical statements. Not only did he choose to pursue 'scientific' philosophy, he also denied that analytical philosophy was of any value.

There is no small irony in Feyerabend plumping for 'scientific philosophy' at this point. His initial conception of philosophy represents a deeply scientistic starting-point for someone whose later work is seen as a searing critique of the claims of science and the very idea of scientific progress.[2] In subsequent early writings, after he had more or less explicitly scouted any distinction between *a priori* and empirical truths, Feyerabend's conception of philosophy settled down into one that was still scientistic, but less vibrantly so, and thus moved closer to that of mainstream contemporary philosophers. But throughout his career Feyerabend railed against analytical and 'linguistic' philosophy, insisting that philosophy should help science and humanity progress, rather than hinder them or simply try to clarify science from the sidelines.

1.2 The Third-Person Approach to Epistemology

Although Feyerabend is thought of primarily as a philosopher of science, his real interest lay in the more general subject of human knowledge, and his approach to that subject was heavily influenced by Popper. To investigate Feyerabend's conception of philosophy of science, we must begin with their intellectual relationship.

Such a suggestion will, however, meet with resistance. The later Feyerabend chastised those who regarded him as a former Popperian. The many acknowledgements to Popper in his earlier papers were, he said, 'friendly gestures, not historical statements' (*SFS*, p. 144n). Complaints about Popper became an oft-repeated theme in Feyerabend's later work.[3] But however strong the current of invective, it is insufficient to dislodge the verdict that Feyerabend was, until the late 1960s, a Popperian.[4] His own work cannot be seen in clear perspective unless we realize that he is tackling the questions 'Is knowledge possible?' or 'How is knowledge possible?', and 'What is the best way to attain knowledge?' These questions comprise just what Popper called the central problem of epistemology, the 'problem of the growth of knowledge' (see Popper [1959], pp. 15–18).

Feyerabend tells us that he discusses scientific theories not out of a wish to restrict himself to philosophy of science (a discipline he often refers to in scathing terms), but because he regards them as 'excellent examples of actual knowledge' ([1965a], p. 217). In his early work he is engaged on the same philosophical project as Popper, the epistemology of science, because during this time he accepted that '*the growth of knowledge can be studied best by studying the growth of scientific knowledge*' (Popper [1959], p. 15). Popper was not alone in his conviction that scientific knowledge is an extension of common-sense knowledge (Russell and Einstein agreed), and that common-sense knowledge grows principally by turning into scientific knowledge. But most importantly he also believed that

> scientific knowledge can be more easily studied than common sense knowledge. For it is *common sense knowledge writ large*, as it were. Its very problems are enlargements of the problems of common sense knowledge. (Popper [1959], p. 22)

To say that Feyerabend is enlisted in pursuing the epistemology of science is not yet to say that he was a Popperian. The project outlined is common to thinkers of many divergent standpoints, few of whom could be described thus. But it does help to explain why, although Feyerabend claims to be searching for a model for the acquisition of knowledge, his abiding concern was with *scientific* knowledge.

This starting-point, however, must be probed. By assimilating common-sense knowledge to theoretical science, it presupposes that general epistemic rationality and scientific rationality coincide. We have to take seriously the possibility that they do not, that they are attuned to different goals. If this is the case, one cannot do general epistemology by doing epistemology of science. If there is no single all-purpose rational method of inquiry, usable in any domain (a later Feyerabendian conclusion which would undercut the rationale for his early work), if science is *not* 'common sense writ large', then the epistemology of science is no more than a corner of epistemology as a whole, and may be of no greater intrinsic importance than any other part, as many of the 'linguistic philosophers', whom Popper and Feyerabend despised, always suspected.

Popper and Feyerabend also shared a conception of how epistemology should be done, a conception Popper came to call 'epistemology without a knowing subject',[5] and which he contrasted with a more traditional conception of epistemology. Adherents of this more traditional conception, such as the British Empiricists, Kant, Mill, Russell, and Husserl focus on 'the knowing subject' and try to figure out its relation to 'the external world'. Third-person epistemology, by contrast, focuses on the products of knowledge: the statements, laws, and theories which cognitive activity issues in. There is nothing exclusively Popperian about third-person epistemology either, since other major twentieth-century philosophers such as Wittgenstein, Quine and Davidson, also take a third-person approach to epistemology. Feyerabend's full Popperian credentials only become clear when we identify him as a contributor to the project of giving a rational model of science, and as an adherent of three crucial views: normative epistemology, falsificationism, and inductive scepticism.

1.3   Feyerabend's Project: A 'Model for the Acquisition of Knowledge'

What we might call[6] a *rational model* of science specifies both an aim for science, and a methodology for approaching that aim. I shall eventually concur with Feyerabend's later suggestion that there is no single such model to be had. But in his early work Feyerabend was as much a contributor to the project of developing a rational model of science as Popper. He characterized the aim of his early work as being 'to present an abstract model for the acquisition of knowledge, to develop its consequences, and to compare these consequences with science' ([1965c]: *PP*1, p. 104). The papers which he wrote between 1957 and the late 1960s, and which will be the major focus of my attention in this book, are most profitably understood as a contribution to this project. A study of Feyerabend's attempt to develop this model is an essential prerequisite for understanding his infamous 1970s work.

We can learn an important thing about Feyerabend's philosophy from his decision to pursue the epistemology of science and from his aim to construct a model of scientific knowledge. This is the overriding importance he attached to the attainment of *knowledge*. Feyerabend took knowledge of scientific statements, rather than understanding, insight, wisdom, conceptual clarity, or enhanced experience, to be the aim of science and therefore of scientific philosophy, and this sets the tone for his early work.

We might also enter some reservations about this project. Science can be said to aim at understanding, explanation, the discovery of natural laws, prediction, and technological control, among other things. Not all of these are profitably characterized as forms of knowledge. Unless we understand 'knowledge' in a loose sense, Feyerabend's position here involves a distortion. Scientific knowledge, narrowly understood as knowledge of scientific propositions, cannot be said to be the exclusive aim of science itself, let alone of philosophy. Just because the epistemologist is primarily interested in science under its aspect as knowledge-gathering does not guarantee that the aims of such a multifarious activity can be characterized in exclusively epistemological terms. The question 'What is the aim of science?' may well be misconceived if it is supposed to have a single answer. The truth may be that science affords us a variety of things, not all of them covered by the concept of knowledge. The idea of a rational model of science is already put in some jeopardy, therefore, by the suspicion that science has no single aim. What is more, as we are about to see, Feyerabend's way of characterizing the methods of science is flawed.

1.4   Normative Epistemology, and Falsificationism

When epistemology is conceived of as a description of the conditions under which knowledge exists, or is attainable, it is said to be descriptive or naturalistic. Two of Feyerabend's most important teachers, Viktor Kraft and Karl Popper, both rejected such a conception. Instead, they conceived of epistemology as a wholly normative discipline, a discipline which lays down rationally grounded rules or norms which, if followed, would produce good science. Feyerabend initially took this conception on board more deeply than any of his contemporaries.

I shall concentrate here on relating the views of Kraft, whom Feyerabend saw as anticipating some ideas now associated with Popper.[7] Kraft investigated the nature of knowledge and the way in which its essential characteristics can be determined, arguing that 'the idea of knowledge can be arrived at only on the basis of stipulations and its validity is a matter of agreement (with these stipulations)'.[8] The theory of knowledge, he thought, must be 'very different from the factual sciences; it does not deal with something existing in reality but puts forth aims and norms for our intellectual activity' (Kraft [1960], p. 32); it suggests a set of ideals which we use to criticize or praise knowledge claims. This bears comparison with Popper's suggestion about how proposed aims of science should be suggested and evaluated. It is the same conception of epistemology Feyerabend committed himself to when he insisted that *the issue between positivism and realism is not a factual issue which can be decided by pointing to certain actually existing things, procedures, forms of language, etc., it is an issue between different ideals of knowledge'* ([1958a]: *PP1*, pp. 33–4). In the same breath, Feyerabend explicitly acknowledged that this thesis is an extension of Popper's views on scientific method, and that the normative character of epistemology had been stressed by Kraft.[9]

Feyerabend praised Kraft's procedure as being the kind of bold and optimistic attempt to change and criticize existing theories, and to develop new theories, which led to modern science in the first place. He followed Popper and Kraft in thinking that the epistemology of science is not a descriptive discipline, but has a normative cast: it tells us what *good* science *should* be. His most explicit statement to this effect is as follows:

[S]cientific method, as well as the rules for reduction and explanation connected with it, is not supposed to describe what scientists are actually doing. Rather, it is supposed to provide us with normative rules which should be followed, and to which actual scientific practice will correspond only more or less closely. It is very important nowadays to defend such a normative interpretation of scientific method and to uphold reasonable demands even if actual scientific practice should proceed along completely different lines. It is important because many contemporary philosophers of science seem to see their task in a very different light. For them actual scientific practice is the material from which they start, and a methodology is considered reasonable only to the extent to which it mirrors such practice. ([1962a], p. 60)[10]

But Feyerabend criticized Kraft for not following up this conception of epistemology with an equally clear conception of the norms to be adopted, and of the reasons for adopting these norms. This, he said, 'would have been a revolutionary undertaking indeed, the first construction of a purely normative epistemology' ([1963c], p. 320). This is what Feyerabend undertook to provide.

Kraft, like Popper, held that almost all the procedures of science and of common sense involve hypotheses, such as the hypothesis of the existence of material things and the hypothesis of the existence of other minds. He thought of these as 'unjustified conjectures for which no foundation can be given' ([1963c], p. 322), but argued that the only alternative to their use is solipsism, the view that the only knowable things are one's own momentary psychological states. We are therefore forced into an uncomfortable choice between solipsism and a body of doctrine which makes essential use of unjustifiable conjectures. Kraft refused to embrace solipsism, and hence resolutely denied that our hypotheses can be justified. But, like Moses, he could not enter the promised land he had seen: he failed to make the transition to *falsificationism*.

A rational model of science, recall, specifies an aim and a methodology for science. If knowledge is the aim of science according to Feyerabend, its methodology, the means of achieving that aim (and the way out of Kraft's dilemma) is falsificationism. Feyerabend suggested a scientific methodology consisting of two principles: the demand that our theories be testable, and the demand that they explain known phenomena. The former ensures that we connect our theories with experience, the latter specifies the connection with experience in detail: our theories must not be *ad hoc*, and they must be richer in content than what they are to explain. It also results in their transcending experience, for 'if the entities postulated for explanation completely coincide with laws in the domain of experience, then they are *ad hoc* with respect to these laws and therefore no longer capable of giving a satisfactory explanation' ([1963c], p. 323). Since theories that do not go beyond experience must be unacceptably *ad hoc*, we should feel no embarrassment about constructing and pursuing theories, even the most highly metaphysical ones, which go well beyond the data. All theories are on the same footing in being *hypotheses*. But this does not mean that we are wrong to theorize:

> The hypothesis of the existence of material objects is . . . not only an essential part of our thinking, in the sciences and within common sense, a part without which much apparently valuable knowledge would simply collapse . . . it is also capable of support. The support is not by *proof*, nor by 'induction' it is by methodological argumentation that is by reference to some of the norms which constitute our epistemology. ([1963c], p. 323)

Such a normative epistemology, Feyerabend thought, can take care of problems (like the 'problem of induction') for which no other satisfactory solution is available. He thus embraced what is now referred to as 'naive falsificationism'.

Kraft, Popper and Feyerabend are all conventionalists about our account of the fundamental nature of science and methodology: they think that there is 'no fact of the matter' about the aim of science, and no fact of the matter about which methodological rules scientists really follow. For them, science has no 'nature'. Instead, we can see science under the aspect of different ideals, each of which will generate a different picture of the scientific enterprise. As philosophers of science, our role is to attune our picture of science not to facts about scientific practice, but to the most worthy set of scientific ideals.

This Popperian conception of the epistemology of science, the single most important root of Feyerabend's philosophy, has been neglected by commentators. But it is not a root which we can leave untouched. Whereas the other problems we have identified so far may seem to represent harmless idealizations, to think of the task of the philosopher of science as primarily normative is to fall into a greater confusion, a conflation of the epistemology of science with scientific methodology itself.

Rules of scientific method do not describe what scientists actually do, but rather lay down what they ought to do. They *are*, therefore, normative and not descriptive. But they cannot be totally independent of actual scientific practice, since it is the activities of scientists themselves which determine the rules of scientific method. So methodology and scientific practice simply cannot 'proceed along completely different lines' (as Feyerabend has it in the quotation from

[1962a], p. 60, above). To recognize that scientific method is normative should not lead us into the error of holding that the theory of scientific method (i.e. the epistemology of science, which Popper confusingly calls 'methodology') is normative too. Those 'contemporary philosophers of science' (Feyerabend mentions only Kuhn) who start from actual scientific practice are not to be taken as denying that there exist methodological rules. Of course, they must accept that, although scientific methodology consists of the rules scientists believe they ought to follow, there are times when scientists fail to follow the rules they themselves profess. And Feyerabend is right to say that one cannot judge a methodology by the extent to which it mirrors practice: a methodology is not a description, so it cannot be impugned for failing to describe scientific practice correctly. A methodology is judged mainly by how useful its results are. Nevertheless Feyerabend clearly makes the confused Popperian assumption that it is up to philosophers to devise (rather than to discover) rules of scientific method. We shall soon see this assumption at work in the very heart of his early philosophy.[11]

This confusion between rules and descriptions, as well as Feyerabend's denial of any *a priori*/empirical distinction, are particularly striking coming from one who insists most vigorously on the importance of distinguishing between nature and convention. In his widely acclaimed book *The Open Society and its Enemies*, Popper argued that humans moved away from the primitive, magical attitude which characterized early societies only when they came to consciousness of the distinction between the natural and the conventional. He insisted that there is a sharp and fundamental distinction between, on the one hand, natural laws, which describe 'a strict, unvarying regularity which either in fact holds in nature or does not hold', and, on the other hand, normative laws, 'such rules as forbid or demand certain modes of conduct' (Popper [1945], p. 57). These Popperian pronouncements are repeatedly referred to in glowing terms by the early Feyerabend. He often cites this chapter of Popper's book (e.g. [1958a]: *PP*1, p. 19, n. 6), which he refers to as containing 'an excellent discussion of [naturalism], its history, and its shortcomings' ([1960c]: *PP*1, p. 234, n. 18), and 'an excellent account of the history of ethical norms and factual descriptions' ([1961a], p. 74). The importance of the discovery of the distinction between nature and convention, he says, 'can hardly be exaggerated' ([1961a], p. 50).[12] But he fails to recognize that as long as he accepts this distinction he must acknowledge the possibility of a distinction between empirical truths and rules (and thus something like the analytic/synthetic distinction).

1.5   Inductive Scepticism

As well as initially accepting a Popperian conception of epistemology and of scientific methodology, Feyerabend endorsed the
Popperian critique of inductive method, and consistently opposed
the position known as *inductivism*. According to inductivism, which
was popular among the Logical Positivists, scientific theories are
empirical generalizations which can be obtained in two steps:
collecting numerous observations of different objects having the
same property P (observations which can be stated in the form 'Pa',
'Pb', 'Pc' etc.), and then inferring, inductively, to a general statement
according to which all objects of the relevant kind have that property
(a statement of the form 'For every object x, x has property P' (in
symbols: '(x)Px')). Like Popper and Kraft, Feyerabend believed that
no such inference can be in the least degree warranted, that the
problem of induction shows that there is no valid method of justification besides deduction. To those who would defend inductivism
with something like the 'analytic' justification of induction, which
says that there can be no proof that inductive inference is unjustified
since to be willing to reason inductively is part of what it *means* to be
rational, Feyerabend sternly responds that 'the standards implied
in common behaviour are themselves open to criticism and . . . it is
the task of the philosopher to provide such criticism, and not to be
satisfied with popularity' ([1964b]: *PP*1, p. 204). In making this
answer, he implicitly appeals, like Popper, to a normative conception of epistemology. But Feyerabend's attitude to the problem of
induction is rather different from that of Popper who, modestly,
claimed to have solved it.

Feyerabend gives his own mini-history of the problem. Originally
it was believed that a finite conjunction of separate true observation-
statements ('Pa' & 'Pb' & 'Pc'. . .'Pn', which we can abbreviate 'P(n)')
could guarantee the truth of the universal conclusion '(x)Px'. This
simple generalization hypothesis was refuted by David Hume, who
pointed out that the falsity of the universal conclusion was fully
compatible with the truth of the conjunctive premise, one negative
instance sufficing to refute the conclusion. So this hypothesis was
replaced by the probabilistic generalization hypothesis, according to
which a conjunction of observation reports could guarantee a high
probability to the universal generalization. Hume's argument, according to Feyerabend, refutes this hypothesis too (he does not say
how). The ongoing search for a hypothesis which would be weak
enough to escape Hume's argument next threw up the modified

generalization hypothesis, according to which given the finite con-junction P(n) it is *reasonable* to infer (x)Px.

Feyerabend complains that this hypothesis, which he sees lying behind almost all recent attempts to solve the problem of induction, including Popper's, is hopeless, partly because what is reasonable or not is 'a notoriously vague affair' ([1964b]: *PP*1, p. 204). But his central complaint is that *all* these hypotheses presuppose an unacceptable move. If there were a solution to the problem of induction, it would be a successful demonstration that there is some justification for asserting the truth of the universal statement '(x)Px', given the truth of the conjunction of singular statements 'P(n)'. The success of this, the modified generalization hypothesis, would, according to Feyerabend, constitute a reason for refusing to consider 'theories' which agree with P(n), but which disagree with (x)Px: 'One almost never starts with P(n) and asks *what* generalization should be adopted. One *takes it for granted* that adopting (x)Px is the right thing to do, and one looks for some plausible argument supporting this belief' ([1964b]: *PP*1, p. 204). In other words, any attempt to justify inductive inference involves a restriction on the number of conclusions one is allowed to consider, and a correlative neglect of alternative conclusions. Feyerabend, who at this time was arguing forcefully for theoretical proliferation (see chapter 7), considered this unacceptable.

In fact, the modified generalization hypothesis, as stated, obvi-ously does not presuppose the unacceptable move. To say that it is reasonable to infer (x)Px is fully compatible with being willing to consider other conclusions. If he really opposed the modified gener-alization hypothesis, Feyerabend would be denying that (x)Px is even *a* reasonable thing to infer from P(n). But he himself thereby would be removing *this* 'theory' from the set of all alternatives, thereby violat-ing his own canon of theoretical proliferation. However, a reasonable inductivist must endorse something substantially *stronger* than the modified generalization hypothesis, something like a rule to the effect that (x)Px is the *most* reasonable thing, or even the *only* reason-able thing, to infer from P(n). Such rules do indeed have the effect Feyerabend is concerned with: they encourage us not to consider some of the other logically possible conclusions.

Feyerabend then reasons as follows. Assume that the modified generalization hypothesis is false, that given P(n) it is *not* reasonable to prefer (x)Px to its alternatives.[13] This would also refute the simple and the probabilistic generalization hypotheses. In fact, it would be even stronger than Hume's disproof, for where Hume showed the impossibility of obtaining the truth or high probability of (x)Px given

P(n), a refutation of the modified generalization hypothesis would also show its undesirability: 'It would not only show that the problem of induction *cannot* be solved; it would also show that it *should not* be solved' ([1964b]: *PP*1, p. 205). Such a refutation, he says,

> demands of us a completely new attitude towards the so-called 'problem of induction'. The fact that the problem is so difficult to solve need not worry us any longer. As a matter of fact, we should rejoice that we are not restricted, by some proof, to the use (given P(n)) of *one* generalization only and are thus able to discover some perhaps decisive shortcomings of this generalization. ([1964b]: *PP*1, p. 206)

Feyerabend's inductive scepticism is thus so radical that it goes beyond both Hume and Popper. The lesson he eventually draws is that good scientific theories not only do not repeat the observational evidence on which they are based, because they go beyond that evidence, but that they actually contradict the evidence. As an example he gives the theory put forward by the pre-Socratic thinker Thales, according to which everything that exists consists of water. Although this might seem to be a case of rank speculation, it is a perfectly good scientific theory, says Feyerabend, since it is both general and explanatory. A theory, like this one, that contradicts the existing evidence, 'provides means for investigating not only certain speculative schemes but even observation reports, which are, after all, not absolutely trustworthy' ([1961a], p. 18). It is therefore a much better instrument of criticism than a theory which leaves things untouched on the observational plane. According to Feyerabend's anti-inductivism, then, theories can take no pride in agreeing with experimental or observational results, for such 'support' is both cheap and flimsy.

1.6   The Ethical Basis of Philosophy

Throughout Feyerabend's early work, the normative conception of epistemology provides the final court of appeal. He always thinks that arguments can be clinched by appealing to the consequences of our 'methodological' decisions. Indeed, one of the most consistent themes in his work as a whole is his strong voluntarism. He insists that the kind of knowledge, science and society we have is up to us, that because epistemology (and therefore philosophy of science) is normative, and because our knowledge depends not on how things are in a world independent of our will but on our decisions, the decisions we make can and must be evaluated by reference to our ideals. Some of these ideals, perhaps the most important ones, will be ethical. This means that the form of our epistemic inheritance can and should be adjusted by reference to ethical norms.[14] Feyerabend is quite explicit that ethics is the basis of epistemology:

> The following fundamental problem: which attitude shall we adopt and which kind of life shall we lead? . . . is the most fundamental problem of all epistemology. . . . [W]e are confronted with a real *decision*, that is, a real choice with a situation which has to be resolved on the basis of our demands and preferences, and which cannot be resolved by proof. It is easy to see that these demands and these preferences concern the welfare of human beings and are therefore ethical demands: epistemology, or the structure of the knowledge we accept, is grounded upon an ethical decision. ([1961a], pp. 55–6)

He even extends the thesis to the whole of philosophy:

> The fact that almost any philosophical doctrine may find realisation either in a *cosmology* . . . and/or in a *theory of man* . . . makes it very clear that the procedure leading to the adoption of a philosophical position cannot be *proof* . . . but must be a *decision* on the basis of preferences . . . Philosophers have habitually judged the situation in a very different manner. For them, only *one* of the many existing positions was true and, therefore, possible. This attitude, of course, considerably restricts the domain of responsible choice . . . [T]he problem of responsible choice enters even the most abstract philosophical matters and . . . ethics is, therefore, the basis of everything else. ([1965a], p. 219 n. 5)

Among the goals that Feyerabend appeals to are: humanity, respect for the individual, happiness, joy, pleasure, imagination, sense of humour, and capriciousness. He often associates these goals with the natural attributes of children, before institutionalized education gets to work on them. Feyerabend sums up the task of his philosophy by saying that a study of science is important because

> despite the great amount of conservatism that is still contained in this enterprise, we have here *criticism* and *progress through revolutions* that leave no stone unturned and no principle unchanged. These changes are not so completely arbitrary and unreasonable as are the changes of ideology following the death of a tyrant or some other 'great man'. They are argued and brought about according to very simple methodological rules. Collecting these rules, eliminating from them the remaining traces of dogmatism and blind ideology would seem to be the starting-point of a type of knowledge that is open to improvement, and therefore *humanised*. ([1965a], p. 217; emphases added)

Insistence that ethics is the basis of philosophy fits with the normative conception of epistemology. But both stand in tension with Feyerabend's conception of philosophy as scientific. The idea that science and scientific philosophy make discoveries about the world presupposes that the world (and therefore, presumably, its constituent objects, events and practices) has a nature that is there to be discovered. The normative conception of epistemology denies that science has such a nature.

2

Meaning: The Attack on Positivism

2.1 Wittgenstein's Conception of Meaning

So far we have exposed one central root of Feyerabend's early philosophy, his Popperian conception of the epistemology of science. Another such root is nourished by Feyerabend's interpretation of the later philosophy of Wittgenstein. While there are other relevant figures in Feyerabend's formative years (Ludwig Boltzmann, Ernst Mach, Viktor Kraft, Felix Ehrenhaft, Philip Frank, Bertholt Brecht, David Bohm etc.), his early work can be understood largely as an attempt to combine the insights of Wittgenstein with those of Popper.

Within semantics (the theory of meaning) there is a broad contrast between 'realist' (or 'representationalist') conceptions of meaning, on the one hand, and what we might call 'instrumentalist' (or 'pragmatist') conceptions, on the other. A realist conception is one in which words or statements are held to have meaning in virtue of their standing for things. An instrumentalist conception, by contrast, asks us to see meaning as a function of the uses to which words and statements are put. In his review of the *Philosophical Investigations*, Feyerabend recognized and explained Wittgenstein's devastating critique of a certain family of theories of meaning, theories associated with 'realism' or 'essentialism'. He interpreted Wittgenstein as reducing such theories to absurdity by showing that they imply that we may never know, of a word we regularly use, what it means, or even whether it has any meaning at all.

My concern here is not with Feyerabend's exposition of this

negative argument, but with the positive conclusions about meaning which he drew from his study of Wittgenstein's text. I shall not defend in detail any of the non-realist theories which Feyerabend ascribes to Wittgenstein, or argue in detail against realist theories of meaning. My aim is merely to investigate Feyerabend's interpretation of Wittgenstein, insofar as this bears on his early work.

In somewhat cavalier fashion, Feyerabend asserts that even though Wittgenstein did not see himself as developing a philosophical theory, everything of interest in the *Philosophical Investigations* comes from treating it as expounding a new theory of meaning.[1] But *which* theory is this? The problem is that Feyerabend attributes to Wittgenstein different and quite incompatible accounts. I shall argue that although Feyerabend at one point did glimpse the nature of Wittgenstein's approach to meaning, in reading his own preferred theory into Wittgenstein's work, he attributed to Wittgenstein the wrong conception entirely.

In his review, Feyerabend ascribes to Wittgenstein 'a new (instrumentalist, nominalist, or whatever you like to call it) *theory of meaning*' ([1955]: *PP2*, p. 125). Discussing the primitive 'language-game' of Wittgenstein's builders (Wittgenstein [1953], § 2), he urges that it suggests 'an instrumentalist theory of language', or

> an intuitionist (pragmatist, constructivist) theory of language – the expressions 'intuitionist' or 'pragmatist' being used in the way in which they serve to describe one of the present tendencies as regards foundations of mathematics. . . . Wittgenstein's theory of language can be understood as a constructivist theory of meaning, i.e. as constructivism applied not only to the meanings of mathematical expressions but to meanings in general. ([1955]: *PP2*, p. 111, n. 12)

There is some tension here between two approaches, not clearly separated by Feyerabend.

On the one hand, this passage, written before 1954, contains a striking anticipation of an influential interpretation according to which Wittgenstein proffers a theory of meaning which identifies the meaning of a statement with its 'assertion conditions', the conditions under which a speaker would be justified in asserting it.[2]

On the other hand, in his choice of the terms 'pragmatist' and 'instrumentalist' Feyerabend seems to mean only that Wittgenstein identifies the meaning of a word with its use, and this is the conception he officially ascribes to Wittgenstein ([1955]: *PP2*, pp. 111, 122). Wittgenstein urges us to 'Let the use of words teach you their meaning' ([1953], p. 220), to 'Look at the sentence as an instrument, and at its sense as its employment' ([1953], § 421). He tells us that 'Language is an instrument. Its concepts are instruments' (§ 569). The non-theoretical, 'instrumentalist' interpretation of Wittgenstein advertised by these remarks has been argued for persuasively in recent years.[3] Feyerabend's suggestion that Wittgenstein pursued such an instrumentalist approach to language was an insight that might have led him towards an adequate conception of the meaning of scientific terms.

Unfortunately, Feyerabend failed to follow up either of these suggestions. He travelled so quickly away from Wittgenstein's instrumentalist conception of meaning that by 1958 he was prepared to affirm that there is a 'clear and unambiguous' distinction between the pragmatic properties of a language and its semantic properties ([1958a]: PP1, p. 19). In subsequent papers and remarks on Wittgenstein, Feyerabend ascribed to him yet another conception of meaning, one which was Feyerabend's own official conception in the early phase of his work, and to which we now turn.

2.2   The Contextual Theory of Meaning

Commentators are divided on the subject of Feyerabend's attitude towards the concept of meaning. On the one hand, sympathizers such as Richard Rorty and Ian Hacking who (like Popper) are sceptical about the importance of meaning, claim that Feyerabend simply was not interested in meaning at all, that he neither had nor needed a 'theory of meaning'.[4] Feyerabend eventually came to identify with this view, which issued in the self-defeating position I have elsewhere called his 'semantic nihilism'.[5]

On the other hand, critics, such as Peter Achinstein, Hilary Putnam and Dudley Shapere, believe that Feyerabend played fast and loose with the concept of meaning because he did have a theory of meaning, but a hopeless one.[6] They think that meaning does matter, but that Feyerabend, Kuhn and other proponents of the 'new fuzziness' in philosophy of science have a half-baked idea of what meaning is. I shall show that these critics are correct in attributing a theory of meaning to Feyerabend, and that the theory in question is inadequate, but I will press objections other than the ones they advanced.

In fact, there is truth in the views of both these groups of commentators, because Feyerabend changed his mind. Before he started saying that meaning does not matter, Feyerabend endorsed what he called 'the contextual theory of meaning':

> [T]he meaning of a term is not an intrinsic property of it but is dependent upon the way in which the term has been incorporated into a theory. . . . Once the contextual theory of meaning has been adopted, there is no reason to confine its application to a single theory, or a single language. ([1962a]: *PP*1, p. 74)

In his review of Hanson's *Patterns of Discovery*, where Feyerabend tries most explicitly to synthesize the insights of Popper with those of Wittgenstein,[7] he refers to 'the pitfalls in the principle that differences of use indicate differences of meaning' ([1960d], p. 250), and accuses Hanson of not taking seriously enough 'the idea (which may be found in Wittgenstein, but which has been held by many nominalists before him) that the meaning of our words is a function of the (theoretical) context in which they occur' (p. 248). In the twentieth century, he claims, 'the contextual theory of meaning has been defended most forcefully by Wittgenstein' ([1962a]: *PP*1, p. 74, n. 68). In the long paper 'Problems of Empiricism' which draws much of his earlier material together, we likewise hear that 'the meaning of every term depends upon the theoretical context in which it occurs. Words do not 'mean' something in isolation: they obtain their meanings by being part of a theoretical system' ([1965a], p. 180; see also p. 184). These are only the most conspicuous of many appearances of the contextual theory of meaning. Feyerabend never developed these hints, but we have every right to assume that the contextual theory of meaning for some time represented his considered opinion.[8]

In evaluating this semantic view, matters are complicated by Feyerabend's unusual views as to what counts as a theory. Most critics home in on the following two characterizations:

> [T]he term 'theory' will be used in a wide sense, including ordinary beliefs (e.g. the belief in the existence of material objects), myths (e.g. the myth of eternal recurrence), religious beliefs, etc. In short, any sufficiently general point of view concerning matter of fact will be termed a 'theory'. ([1965a], p. 219, n. 3)

> When speaking of *theories* I shall include myths, political ideas, religious systems, and I shall demand that a point of view so named be applicable to at least some aspects of everything there is. The general theory of relativity is a theory in this sense, 'all ravens are black' is not. ([1965c]: *PP*1, p. 105, n. 5)

The first is intolerably vague, but the second is worse, since any empirical generalization of the form 'All A's are B' is 'applicable to everything there is', being restatable in the form 'Everything is either an A or is not-B'. However, Feyerabend does, in one place, give a more useful definition of theory:

[T]he usual distinction will be drawn between *empirical generalisations*, on the one side, and *theories*, on the other. Empirical generalisations are statements, such as 'All A's are B's' (the A's and B's are not necessarily observational entities [*sic*]), which are tested by inspection of instances (the A's). Universal theories, such as Newton's theory of gravitation, are not tested in this manner. Roughly speaking their test consists of two steps: (1) derivation, with the help of suitable boundary conditions, of empirical generalisations; and (2) tests, in the manner indicated above, of these generalisations. One should not be misled by the fact that universal theories, too, can be (and usually are) put in the form 'All A's are B's'; for, whereas, in the case of generalisations, this form reflects the test procedure in a very direct way, such an immediate relation between the form and the test procedures does not obtain in the case of theories. ([1962a]: *PP*1, p. 44, n. 1)

This more sophisticated conception respects the important point, familiar from the work of Pierre Duhem, that scientific theories are not themselves empirical generalizations, which can be falsified by a single negative instance, but rather are more complex structures which can only be tested together with auxiliary hypotheses in a theoretical system, from which testable empirical generalizations can be derived.

Knowing what counts as a theory and what counts as part of the theoretical context of a term, we now know, according to the contextual theory of meaning, how to determine the meaning of a term.

Commentators have noticed the connection between the contextual theory of meaning and the 'law-cluster' theory adopted by the Logical Positivists, according to which the meaning of a theoretical term in a given theory is a function of the role that term plays in the theory.[9] The mere fact that such an account was held by positivists does not show that it is positivistic in any deeper sense. But the contextual theory is positivistic in virtue of being verificationist: it insists that meaning must be recognizable, that the meaning of an expression cannot transcend our theories (which themselves must necessarily be intelligible to us). Such a conception rules out meaning-scepticism, the worry that our words could have meanings which we are wholly unaware of. But it is correct to do so: one valuable lesson Feyerabend drew from Wittgenstein's work was that the fact that some realist conceptions of meaning leave open the possibility of meaning-scepticism should be regarded as a decisive objection to them. Verificationism about the constituents of the world seems wildly implausible: the way things are in the world may be very different from the way we think they are. Scepticism about claims that we know the way things are makes sense. But scepticism about meaning is ridiculous: the meaning of words and statements in human languages is determined by humans, and must be humanly and publicly knowable. The idea that meaning cannot be verification-transcendent has not been demolished. This verificationist flavour to the contextual theory of meaning does not, however, mean that the theory is in conflict with scientific realism (that is, realism about scientific *theories*): we shall soon see that Feyerabend thought them intimately connected.

What about Feyerabend's claim that it was Wittgenstein who put forward the contextual theory, that he was 'one of the most eloquent defenders of this principle in recent philosophical thinking' ([1965a], p. 252 n. 137)? Here caution is needed. To decide whether a contextual account coincides with the instrumentalist conception of language that Wittgenstein held we need to know what notion of 'context' Feyerabend works with. The term 'context' may, after all, just be intended to refer to a term's context of use. But we have already seen that the way in which Feyerabend explains his contextual theory makes it plain that it does *not* coincide with the instrumentalist conception we identified earlier. The contextual theory states that the meaning of a term or statement is determined by the surrounding context of theoretical principles, syntactic and semantic rules in which it figures. It is important to remember that the context invoked is truly theoretical context and not just any (psychological, sociological) context. Feyerabend did once insist on this: 'I would be reluctant to define so widely the context to which we must refer when explaining the meaning of a scientific term that personal or group idiosyncrasies appear as part of the definition of meaning' ([1960d], pp. 249–50). A contextual theory does not make meaning inherently subjective. What determines the meaning of a term might, after all, be the context of law-like generalizations in which it occurs.[10] But Feyerabend erred in extending the contextual theory, originally an account of the meaning of theoretical terms in science, to apply to all terms. This wrongly presupposes that it makes sense to speak of any term as having a theoretical context, and Wittgenstein should not be implicated in this extension. In extending the contextual theory of meaning to all terms, Feyerabend opted for a monolithic conception of meaning. He forgot that one of Wittgenstein's central aims was to show that not all terms function in the same way (the toolbox analogy in § 11 of *Philosophical Investigations*). Over-extension of the contextual theory ignores the fact that different concepts have different *application conditions* or *criteria*. As long as we do not understand the notion of 'context' as theoretical context, a contextual 'theory' is acceptable as a slogan. As an account of the meaning of *theoretical* terms, the contextual theory was a promising, if oversimplified, first stab.[11] A genuinely theoretical term, which purports to pick out an unobservable object, process or property, is applicable only on the basis of sophisticated criteria. In particular, the inferential connections between such a term and other theoretical terms are crucial. To know the meaning of a term whose primary occurrences are within a scientific theory it is necessary (if not sufficient) to understand its role within that theory, and thus to understand the theory. Note, however, that knowledge of the term's role in a theory must include the ability to apply the theory to empirical phenomena: one cannot be said to understand a theory unless one knows (at least roughly) how it bears on experiment and observation. So a crucial semantic connection of theoretical terms to observation still exists. Nevertheless, theoretical terms are not reducible to observation terms, since an explanation of the extra-theoretic connections of a theoretical term (its connection with observables, such as pointer-readings) does not amount to an explanation of the intra-theoretic role of the term, and explaining this is at least a necessary condition of explaining its meaning.

Most of the terms we use, however, are not theoretical terms. To look for the theoretical context of terms like 'blue', 'shoe' and 'or' is misguided. The meaning of ordinary terms like these is largely determined by their criteria of application, and any connections with other terms are secondary. The use of these terms makes it amply clear that their meaning is not fixed by anything rightly called 'theory', and that any applications they may have in theoretical contexts are parasitic on their use in ordinary contexts. Colour terms and sensation terms are applied non-inferentially, and not on the basis of any grounds or criteria at all. Terms for familiar 'medium-sized specimens of dry goods' *are* applied on the basis of criteria, but it is essential that the criteria are familiar ones, not open to falsification by sophisticated scientific theory.

No monolithic conception of meaning can do justice to the complexity and variety of linguistic phenomena. A monolithic conception asks us to consider those phenomena in the light of a single theoretical concept. The 'instrumentalist' conception of meaning is not to be understood as monolithic: the concept of 'use' is not a theoretical concept in terms of which the meaning of an utterance can be calculated. Rather, this conception forces us to recognize that the issue of meaning is most profitably approached through the internal relations among meaning, use, understanding, and explanation: meaning is what an explanation of meaning is an explanation of; meaning is what is transmitted in the teaching of, and grasped in the learning of meaning, and what is manifested in understanding. Questions of 'content' impinge in a variety of ways and degrees on the question of meaning, but these humble conceptual truths and their ramifications have deep implications for semantics.

We shall see in chapter 6 that Feyerabend's contextual theory of meaning gets him into all sorts of trouble.[12]

2.3   The Contextual Theory of Meaning and Scientific Realism

Feyerabend spent little time arguing for the contextual theory of meaning. Sometimes he just appealed to the authority of Wittgenstein. It appeared in his early writings as an unquestioned tenet because in his most important early paper, 'An Attempt at a Realistic Interpretation of Experience', he *did* argue for the related thesis that '[t]he *interpretation of an observation language is determined by the theories which we use to explain what we observe, and it changes as soon as those theories change*' ([1958a]: *PP*1, p. 31). This idea, which he calls 'Thesis I', and which he explicitly ascribes to Wittgenstein, as well as to Galileo and other scientists, is the core of Feyerabend's version of scientific realism. The contextual theory of meaning, being more general than Thesis I in applying to *any* kind of language, logically implies this thesis. In arguing for this kind of scientific realism Feyerabend may have seen himself as simultaneously providing support for the contextual theory of meaning.

2.4   The Positivist/Realist Dispute

Feyerabend's central philosophical concern was the relation between theory and experience. He thus inherited from his teachers Kraft, Hollitscher and Popper an interest in the most all-pervasive dispute in twentieth-century epistemology of science, the dispute between positivism and scientific realism. Although Feyerabend described himself as having been 'a raving positivist' (*SFS*, p. 112; see also *KT*, p. 72) while he was a science student during the late 1940s, it was Hollitscher, he claimed, who persuaded him of the cogency of realism about the 'external world' (Popper's important arguments for realism and against positivism came somewhat later). Hollitscher's argument was simply that scientific research is conducted under the assumption of realism, and could not be otherwise conducted. Feyerabend developed this thought in a fascinating series of papers beginning in 1957, arguing that science needs realism in order to progress, and that positivism would stultify such progress.

Positivism does not lend itself to easy definition, since new varieties often challenge what might be regarded as the last generation's essential tenets. But we can accept Ian Hacking's 'family resemblance' account:[13] positivists are empiricists who typically emphasize verification, falsification and observation; they feel uneasy about causes, explanations, theoretical entities and metaphysics. Their opponents, scientific realists, want to think of theoretical entities as real, of causal relations as necessary, and of metaphysics as acceptable proto-scientific speculation. They hope to show that a successful theory would actually describe or picture reality. Thus realists have few qualms about the unobservable, and typically emphasize concepts like truth and reference, which they suppose afford us the means of bridging the gap between language, or the mind, and reality itself.

The nature of the positivist/realist dispute has been variously understood since its inception. Feyerabend's distinctive conception, fully consonant with his normative conception of epistemology, is that the dispute concerns which view of science is more conducive to the realization of certain shared epistemic ideals:

> [T]he issue between positivism and realism is not a factual issue which can be decided by pointing to certain actually existing things, procedures, forms of language, etc., it is an issue between different ideals of knowledge. ([1958a]: PP1, pp. 33–4)

Human knowledge, he says, can not only be presented in different forms, each attuned to different ideals, it can also take different forms: we can actually have 'absolute truth', certainty and infallibility if we want. We are free to interpret science in any way we choose, as long as we are prepared to accept the consequences of our choice. If, like positivists, we see knowledge as a systematization of our experiences, we will value certainty, security, and absolute truth, and will try to maximize the secured empirical content of our beliefs. If, however, we place a higher value on potential informativeness (empirical content, as other philosophers conceive it), we will follow realists in seeing our theories as bold conjectures which, in being about a mind-independent world, go beyond our experiences. The one combination we cannot have is certainty *and* informativeness: there is, as Popper emphasized, a strict trade-off between the two.[14]

This room for decision in the dispute between realism and positivism does not make the rejection of positivism arbitrary since, Feyerabend says, 'we judge an ideal by the consequences which its realisation may or may not imply' (*PP*1, p. 35). In his early work the consequences by reference to which we decide in favour of positivism or realism are the consequences for knowledge. We decide by reference to shared epistemic ideals, like the ideal of scientific progress, to which Feyerabend repeatedly appeals. But if, say, the realist could show that positivism is not conducive to the progress of science, then the positivist can still stick to his or her guns by renouncing this ideal; there is no question of showing that positivism is factually or conceptually incorrect.

We might find that, on a positivist account of how scientists use theories, laws and hypotheses, certain aspects of scientific practice seem unintelligible. Such a discovery, however, which would tell conclusively against positivism on a descriptive conception of philosophy, does not conclude the debate as Feyerabend sees it. For him, there is still the further question as to whether the recalcitrant aspect of scientific practice is desirable or not. If it does not contribute towards our ideals, we are at liberty to decide to see science as free from it, and even to construct sciences without it. So when Feyerabend announces that he intends to show that positivism is incompatible with 'scientific method and reasonable philosophy' (*PP*1, p. 17), it is primarily the latter conflict that concerns him. Any disagreement between a philosophical ideal and actual scientific method merely shifts the issue onto the desirability of the method in question. One who does not believe that there are facts about scientific method can hardly chastise positivism for failing to take account of such facts. Working within this normative philosophical project, Feyerabend claims to find an *a priori* argument for realism, which we will outline in chapter 4.

Feyerabend begins his critique of positivism by attacking positivistic interpretations of science, according to which the task of science is the systematization and extension of human experience. He notes that there is in science a rough distinction between theory and observation which is best explained by formulating the conditions which a language must satisfy in order to be an acceptable observation-language. We are asked to suppose that an observation-language L consists of (a) a class of uninterpreted observation-sentences, and (b) an 'interpretation'. The members of class (a) must be such that a group of observers, when presented with any such sentence in an appropriate situation, come to a quick, unanimous, and relevant decision as to its truth-value. (Thus, we might call them 'easily decidable'.) This condition is purely pragmatic, since it concerns features of the use of observation sentences. The pragmatic properties of L are fully characterized by what Feyerabend calls the 'characteristic' of that language, this being a five-place relation the terms of which denote features which determine the use of L's logically non-compound ('atomic') sentences. To qualify as a fully-fledged language, L must also meet the class of further conditions which comprise (b), the 'interpretation'. The interpretation of an observation-language is a semantic element which confers meaning upon its sentences, thus transforming them into observation-statements. A given observation-language is said to be 'completely specified' by its characteristic plus its interpretation. Feyerabend situates the positivist immediately:

> Any philosopher who holds that scientific theories and other general assumptions are nothing but convenient means for the systematisation of the data of our experience is thereby committed to the view (which I shall call the stability thesis) that *interpretations* (in the sense explained above) *do not depend upon the status of our theoretical knowledge*. ([1958a]: *PP*1, p. 20; see also p. 17)

Feyerabend considers Wittgenstein's emphasis on the use of expressions inconsistent with positivistic accounts of science, according to which, 'the interpretation of certain expressions (the descriptive expressions of the "observation-language") is fixed once and forever and is the basis on which all the other expressions obtain their respective meanings' ([1960d], p. 247). Unfortunately, Feyerabend does not distinguish between weak and strong ways of interpreting this 'stability-thesis'. On a weak interpretation, it says only that the meaning of observation-statements *can* be constant across theoretical change. An altogether more radical version, expressing 'the myth of the given', has it that the meaning of observation-statements *must* remain constant. Wittgenstein was not committed to this radical stability-thesis, since he could allow that the use, and thus the meaning, of observation-language sentences can change. But is even the positivist so committed? We shall soon see that a positivist who endorses a pragmatic theory of meaning can accept that there are changes in the meaning of observation-language statements, and thus that the radical stability-thesis is not, as Feyerabend thinks, an unavoidable but undesirable consequence of positivism. Feyerabend's own attempt to show that the positivist must accept the stability-thesis ([1958a]: *PP*1, p. 20, n. 7), being an inductive generalization from two instances, leaves something to be desired.

2.5   Positivistic Theories of Meaning

Feyerabend enters the positivist/realist dispute by tackling the question 'How is the meaning (that is, the "interpretation") of observation-statements determined?' The positivist has two suggestions, one of which Feyerabend calls the 'principle of phenomenological meaning', and which says that 'the interpretation of an observational term is determined by what is "given" (or "immediately given") immediately before either the acceptance or the rejection of any observation sentence containing that term' ([1958a]: *PP*1, pp. 21–2). This represents the core of a sense-datum epistemology, which Feyerabend deems the most plausible way of defending the idea that observation-statements are incorrigible. According to such an epistemology, to explain the meaning of an observation term like 'red' one needs only to create circumstances in which the learner experiences redness. What is 'given' in experience determines the meaning of the word.

Arguments familiar from Wittgenstein's *Philosophical Investigations* show that this kind of approach to meaning is hopeless. It conceives of what is experienced as a logically private object, something to which only one person can possibly have mental access. But logically private objects are not good candidates for what determines the meaning of words.[15] Feyerabend cannot borrow Wittgenstein's arguments against such a 'private language', since he holds that we can decide to introduce just such a language. But he adduces his own objections to the principle of phenomenological meaning.

His main objection is a *reductio ad absurdum*. Consider the relation between an 'immediately given' phenomenon P (e.g. a red patch in one's visual field) and the acceptance of a sentence S (e.g. 'I see a red patch') whose meaning is supposed to be determined by the presence of that phenomenon. Call this the relation of phenomenological adequacy. This relation cannot be 'immediately given' in the same sense in which P is, for then the observer would have to identify the relation between P and S as the relation of phenomenological adequacy. He could have done this only if he compared this latter relation, call it P', with a further phenomenon S' (a thought or sentence) to the effect that P' is the relation of phenomenological adequacy, and by discovering that S' fits P'. But this discovery presupposes further that he attends to the relation between P' and S' and that he identifies this relation as the relation of phenomenological adequacy. Thus we are started on an infinite regress. Therefore S can be false; it can mischaracterize P.

Elie Zahar, commenting upon this argument, claims that Feyerabend relies upon the gratuitous assumption that P', in order to exist at all, must be adequately described by some S' (Zahar [1982], p. 398). The question is whether we can know that S 'fits' P' without having any means to express this knowledge. Zahar finds it probable that 'our phenomenological domain transcends the bounds of every natural language' (p. 399). Such a supposition may be acceptable to a positivist keen on the principle of phenomenological meaning (even though reliance upon a non-linguistic intuition to characterize the relation between P and S as the relation of phenomenological adequacy would *normally* fit ill within a positivist epistemology). But it is hard to see how a meaningful sentence expressing a relationship could have its meaning completely determined by a relationship which cannot be adequately described, and is inexpressible. Such a sentence would be one whose meaning was humanly unstatable: its meaning could therefore be conveyed only in an appropriate situation in which the audience was confronted with the relationship. But how, in such a situation, would the communicator direct the audience's attention to, or know whether the audience had focused on, the correct relationship? Zahar's suggestion seems to involve some of the problems associated with logically private languages. There are two related errors in this line of thought. The first is to conceive of experiences as logically private. The second is to think that experiences (however conceived) determine the meanings of words. I believe, therefore, that Feyerabend's main argument stands. Questions about the meaning of an observation-statement cannot be decided by introspection on the contents of experience. In fact, whether or not one finds Wittgenstein's arguments against the possibility of a logically private language compelling, Zahar's proposal, in representing the relationship between statement and phenomenon as ineffable, injects mysticism into the foundations of empirical science.

A second objection Feyerabend makes is that a sense-datum proposition is incorrigible for at most one person, and is so only at the time of its utterance. Another observer can only conjecture that she is attaching the same meaning to the terms of the original observation report, but without being able to establish this conjecture. Such a complaint is now a familiar objection to any 'phenomenological' theory of meaning, since such theories are incapable of accounting for the *intersubjective* nature of language.

The other positivist suggestion, which Feyerabend calls the 'principle of pragmatic meaning', states that the interpretation of an expression is determined by its use. This is taken to mean that 'the interpretation of an observation language is uniquely and completely determined by its characteristic' ([1958a]: *PP*1, p. 21). The idea that the 'interpretation' of an observation-language is determined by its use, together with the (contingent) fact that this use is fairly stable over time, allegedly implies the stability-thesis (even though, as we saw above, Feyerabend denies that Wittgenstein's use of theory of meaning is so committed). But Feyerabend undertakes to show that, contrary to the principle of pragmatic meaning, 'it is possible for the interpretation of a language to change without any perceptible effect upon its characteristic' (*PP*1, p. 22), that is, roughly, he undertakes to demonstrate that meaning is not determined solely by use, in this, his own formalized sense of 'use'.

2.6 Feyerabend's Attack on the Stability-Thesis

The principles of pragmatic and phenomenological meaning, while associated with positivism, are really only Feyerabend's secondary targets. His main target is the radical version of the 'stability-thesis' which says that the meaning of observation-statements must be constant across theoretical change. Feyerabend's attempt to refute this is also his main attempt to refute the principle of pragmatic meaning.

To do so, he would have us imagine a language L which contains colour predicates ascribable to self-luminescent objects. We are to assume that the 'characteristic' of L, the use of its predicates, is defined, and that the methods of observation involved in determining this characteristic involve only the kinds of low velocities and masses met with every day. The predicates of L are supposed to designate properties of the objects which those objects possess whether they are observed or not. Now imagine that someone proposes a new theory according to which the wavelength of light depends upon the relative velocities of the observer and the light source. Adopting this theory could, according to Feyerabend, lead to a change in the 'interpretation' of L without any change in its characteristic. A statement of L, such as 'Object *a* has colour P', uttered before the adoption of the new theory, asserts that P is a property of *a*. But after the adoption of the new theory this same statement, whose use does not change, is 'no longer complete and unambiguous. It will depend upon a parameter p (the relative velocity of a and the co-ordinate system of the observer . . .)' (*PP1*, p. 30). The new theory asserts that a colour ascription is a *relational* statement (of the form '$P(a,p)$') and not a simple predication. The observer will continue to use the sentence 'Object a has colour P' as he did before when making reports of phenomena on the everyday level; nevertheless his statement does not possess its former 'interpretation'.

In this simple argument we can discern many of the foundations of Feyerabend's philosophy (the contextual theory of meaning, the idea that languages contain theories, and the presumption of incompatibility between non-isomorphic theories), as well as the prototype of his later accounts of meaning-change. If the argument works it impugns any version of the stability-thesis. But for it to work there must be some support for the idea that a change in meaning can occur in the absence of any change in application. As far as I can see, Feyerabend offers no such support. The main problem is that he has not told us what the 'interpretation' of a language (or statement) consists in.

The principle of pragmatic meaning says that meaning is determined by use; Feyerabend took this to mean that the interpretation of a language is determined by its characteristic. In his example, has any feature of the use of the sentence 'Object a has colour P' changed? Let us start by assuming that no such change in the characteristic has taken place, that the sentences of L have exactly the same use they had before the change. What, in this case, could make us think that the meaning ('interpretation') of that sentence has changed? What makes us think that an utterance of this sentence by someone who has accepted the new theory, despite its surface form, must now designate a relation? *Ex hypothesi*, no such change in meaning turns up in a difference in the use of the sentence. In fact, if there was such a change in meaning, it must be one which transcends the speaker's *knowledge* of meaning, for he or she would still explain the sentence in the very same way as before. Feyerabend's complaint therefore licenses the possibility that the speaker should not know what he means by the sentence, despite his being fully competent in its use. But this is exactly the consequence of 'realist' semantic theories that Feyerabend, following Wittgenstein, considered unacceptable. The defender of the principle of pragmatic meaning will therefore retort that the sentence has suffered no change in meaning, that it simply does not now mean '$P(a,p)$'.

If, instead, we assume that the characteristic of the language has changed, then defenders of the pragmatic principle can point to changes in the (possible, if not actual) correct use of the sentence which warrant our thinking that it has suffered a change in meaning. Feyerabend's example thus does not succeed in illustrating either how the 'interpretation' of a language can change without any perceptible effect upon its characteristic, or how that interpretation depends upon our theoretical knowledge. Neither does it refute any reasonable 'stability-thesis' positivists might endorse. They need be committed only to the idea that the meaning of observation terms can be stable across theoretical transitions, not that it must be. Of the two semantic theories Feyerabend offers them, only the principle of pragmatic meaning allows for the possibility of meaning-change. Reasonable positivists should therefore endorse the principle of pragmatic meaning, not the principle of phenomenological meaning.

Apart from his attempt to refute it outright, Feyerabend tries to show that the stability-thesis has consequences unacceptable to positivists. We make assumptions, he holds, not only in assertion, but also merely by using a language as a means of communication. Any statement implied by the assertion that a language L is applicable he calls an 'ontological consequence' of L. The stability-thesis, he says, implies the existence of ontological consequences which are not logically true, for,

> assume (a) that the observation language has ontological consequences; (b) that it satisfies the stability-thesis . . .; and (c) that it is applicable, was applicable and will always be applicable. Then it follows, (1) that those ontological consequences cannot have emerged as the result of empirical research (for if this were the case, the stability-thesis would have been violated at some time in the past); (2) nor will it ever be possible to show by empirical research that they are incorrect (for if this were the case the stability-thesis would be violated at some time in the future). Hence, if the ontological consequences of a given language are not all logically true statements. . . *every positivistic observation language is based upon a metaphysical ontology.* ([1958a]: *PP*1, p. 21)

Such a consequence would be unacceptable to positivists, who are allergic to metaphysics. But the only reason Feyerabend gives why the ontological consequences of an observation-language should not be logically true is that if they were, that language would be applicable for purely logical reasons, and this, he says, is implausible. If we discover that such ontological consequences are false, this amounts to a discovery that such a language is 'inapplicable to reality'.

This is a strange objection. Languages are, by definition, 'applicable to reality', and logically-true statements cannot be discovered to be false. Feyerabend errs here in assimilating languages to theories, thinking that some languages 'fit' reality better than others, and in thinking that a change in whether we accept a sentence which expresses a logical truth amounts to its falsification.

Another consequence of positivism is supposed to be that the stability-thesis will give a correct account of the interpretation of our observation-language. This will lead to a metaphysical ontology, the basis of which is a theory conserved because it appears to be phenomenologically adequate. But the price to be paid is that this theory will be completely void of empirical content, and there will be a diminution, perhaps even a total loss, of the critical, argumentative function of our language: '[P]ositivistic knowledge', declares Feyerabend, 'is connected with a more primitive and naturalistic stage of human development than its alternative' ([1958a]: *PP1*, p. 35). This objection involves an appeal to the 'methodological principles' of empiricism and testability, which will be considered later.

3
Theories of Observation

3.1 The Theory-Ladenness of Observation-Statements

According to 'Thesis I', which encapsulates Feyerabend's version of
scientific realism, the interpretation of an observation-language is
determined by the theories which we use to explain what we observe.
Feyerabend claims that this thesis should be judged by the plausibility of its consequences, among which are that

> [t]he distinction between observational terms and theoretical terms is a
> pragmatic (psychological) distinction which has nothing to do with the
> logical status of the two kinds of term. On the contrary, thesis I implies that
> the terms of a theory and the terms of an observational language used for
> tests of that theory give rise to exactly the same logical (ontological)
> problems. *There is no special 'problem of theoretical entities'*. ([1958a]: *PP*1,
> p. 32)

This chapter examines these consequences of Feyerabend's realism.
   Feyerabend is well known for believing in the 'theory-ladenness of
observation', but he was by no means the first philosopher to advance
such an idea. Let us distinguish between the following importantly
different positions:

1 Observation is 'theory-laden' in that it is shaped by prior knowledge of the scene observed. (Hanson [1958], p. 19)
2 Scientists use, in their observation reports and elsewhere, 'theory-
   loaded' nouns, words which only make sense against a certain
   background. Such words are sometimes used in a 'theory-loaded'

fashion and sometimes are used as 'data words'. Which are the 'data words' and which the 'theory words' depends on the context. (Hanson [1958], pp. 54ff)

3 No predicates, not even those of an observation-language, can function by means of direct empirical associations alone. All descriptive predicates are theory-dependent in that 'the functioning of every predicate is found to depend essentially on some laws or other', where a 'law' is a general statement accepted as true. (Hesse [1974], pp. 11–16)

4 Observation-statements are not merely theory-laden, but *fully theoretical*; they have *no* 'observational core' (*PP1*, p. x).[1]

Deciding which of these positions Feyerabend accepted is attended with difficulties. Feyerabend should probably not be represented as believing consistently in the theory-ladenness of observations (position 1), for two reasons. First, although he did initially accept that view, '[h]aving been influenced by Wittgenstein, Hanson and others' (*AM1*, p. 133; *AM3*, p. 96), he soon came to think it empirically refuted. The view that we see things as we believe them to be, or that every major conceptual change will affect our perceptions, he says,

> has . . . been held without experimental support. I strongly sympathise with this latter view, but I must now regretfully admit that it is incorrect. Experiments have shown that not every belief leaves its trace in the perceptual world and that some fundamental ideas may be held without any effect upon perception. ([1965c]: *PP1*, p. 128)[2]

Second, as we shall soon see, Feyerabend's own theory of observation actually requires that theories be tested against experience, and it therefore needs the theory-neutral substratum which he calls 'human experience as an actually existing process' ([1965a], p. 214). His theory of observation says that our experience causes us to utter sentences, and our theories are acceptable to the extent that they reproduce these utterances. If experience *itself* were theory-laden, this kind of theory-comparison could not happen. Dudley Shapere drew the correct conclusion, that '[i]t is not theory-independent observation, but a theory-independent observation *language* that Feyerabend is set against' (Shapere [1966], in Hacking [1981], p. 47). Still, this leaves us three theses about the theory-ladenness of language to decide between.

Gilbert Ryle, one of the Oxford 'linguistic philosophers' Feyerabend had little time for, introduced position (2).[3] In his review of the book in which Hanson endorses this view, Feyerabend suggested that it is an immediate consequence of the contextual theory of meaning. Even so, he never showed a great deal of interest in this position, presumably because he did not accept that theory-ladenness holds only of a particular class of words or uses of words. Ryle and Hanson's view that some terms and some uses of terms are 'theory-loaded' correctly presupposes that others are not. But Feyerabend cannot take advantage of it.

Feyerabend certainly espoused position (4). He later claimed that this had always been his view, and that he had never held (3) at all:

> Theory-ladenness means that there is a theoretical load and something non-theoretical that carries the load contained in every observation statement. I have *opposed* this thesis in *all* my writings. (*SFS*, p. 157, n. 9; emphasis added)

But (4) is impossible to reconcile with some of the things Feyerabend says about the meaning of observation-statements, such as the following:

> [T]he 'logic' of observational terms is not exhausted by the procedures which are connected with their application 'on the basis of observation'. . . . it also depends on the more general ideas that determine the 'ontology' (in Quine's sense) of our discourse. These general ideas may change without any change of observational procedures being implied. ([1963a], p. 16)

This fits better with position (3), where that is understood as allowing that a *part* of the meaning of an observation-statement or term is determined either by its use or by the phenomenological situation which allegedly must attend its introduction.

The truth is probably that Feyerabend switched from (3) to (4) for 'methodological' reasons. Whereas in 1963 he was prepared to allow that part of the meaning of observation-statements is determined by use or phenomenology, by 1965 he was claiming that observation-statements have, or at least should have, no 'observational core'. Sensations, he argued, need interpretation. Interpretations are chosen by us. If observation-statements have an observational core, this will be a result of our own stipulation: we will have given them this kind of content. But this would be, methodologically speaking, a bad move. If we endorse a reasonable methodology, the interpretation we choose 'must make impossible the perennial retention of factual statements', and this condition 'eliminates the idea of an observational core' ([1965a], p. 216). Observation statements must be interpreted wholly in terms of theories.

Feyerabend's line of thought here is pretty clearly mistaken. To reject the idea that there are factual statements which can be retained through any change in belief is to give up the idea of incorrigible observation-statements. But this is not necessarily to accept that factual statements have no sensory content, no core of observational meaning whatever. It is, rather, to say that we cannot linguistically isolate this component of meaning by producing statements which express only what is 'given' in experience. It is hard to see how statements which have *no* observational meaning could possibly count as observation-statements. This is why critics accused Feyerabend of denying the existence of observation-statements, rather than giving an account of their nature.

What about the more moderate position, (3)? Anyone not enamoured of phenomenological theories of meaning will concede that no predicates can function by means of direct empirical associations alone. But this does not, as Feyerabend thinks, force one to accept that all observation-statements are theory-laden. We have to inquire after the status of the accepted general statements whose truth provides the background for the correct application of observation predicates. Some of these general statements are not laws at all, but rules for the correct application of the predicate in question. These constitute not a theory but rather a concept of whatever the predicate picks out. They should not be represented as falsifiable, that is, as liable to fall foul of advancing empirical information. Of course, this does not mean that they cannot be altered. It means that their alteration amounts to conceptual, rather than theoretical change, a change in meaning rather than a change in belief.

Feyerabendians, who scout any distinction between theoretical and conceptual change, are not impressed by this objection. They also posit cases where indisputably empirical information is prerequisite for the application of observation predicates. Perhaps there are such cases. But there is still the question whether this collateral information, the need for which ensures that observation-statements are not incorrigible, amounts to theory, properly so-called. Often the information required is just too superficial and humdrum to count as theory.[4] Needless to say, Feyerabendians, whose conception of theory embraces *any* information-bearing cognitive states, do not appreciate this reservation either.

Whether Feyerabend held (3) or (4), his view implies that we cannot divide the language of science, or any language, into a theoretical and a non-theoretical part. The meaning of observation terms and observation-statements depends on the postulates of the theories in terms of which they are interpreted. If this is true, any positivist who

presupposes the stability of such meanings is wrong. But contemporary 'positivists', in the shape of logical empiricist philosophers such
as Herbert Feigl and Ernest Nagel, soon objected that Feyerabend's
theory-ladenness thesis makes nonsense of the idea of *crucial experiments*.[5] A crucial experiment is one used to decide between two
competing theories. It requires that there be some observation-statement which is affirmed by one of the theories and denied by the other.
To be relevant to both theories, any such observation-statement must
have the same meaning in both. Feyerabend allowed that crucial
experiments which decide between low-level theories make sense. In
such cases, he argued, background theories provide a common
interpretation for the observation-statements of both theories, 'and
thereby make crucial experiments possible' ([1965a], p. 216).

It is with his favourite 'high-level' or 'global' theories that the
procedure breaks down. Nagel noted that on Feyerabend's view
observation-statements are doubly theory-laden: they go beyond the
information 'given' in experience, so that they can never be completely verified; and the meanings of the terms they contain can be
explained only by explaining the theories that inform them. He then
argued that there *must* be observation-statements whose meanings
are neutral with respect to the theory being tested:

> [S]uppose that some theory is being tested by obtaining certain experimen-
> tal data. It would clearly be circular to interpret those data in such a way
> that they are described with the help of terms whose definitions presup-
> pose the truth of the theory. Accordingly, observation statements contain-
> ing such terms can be used in this case, only on pain of making the theory
> irrefutable, and thereby depriving it of empirical content. (Nagel [1979],
> p. 79)

For the case of high-level theories, Feyerabend urged us to take his
'pragmatic theory of observation' seriously and to 'accept the theory
whose observation sentences most successfully mimic our own behaviour' ([1965a], p. 217). This theory of observation we are about to
examine.

Apart from recognizing the problems which Feyerabend's thesis of
theory-ladenness generates, we should also note the resources which
it affords him. Feyerabend believed that evidence derived from
observation, although relevant for assessing scientific theories, deserves nothing like the decisive weight that positivists, falsificationists
and logical empiricists give it. For him, the support which positive
empirical evidence lends theories is both weak and cheap, and the
obstacles that negative evidence puts in the way of theories are

ineffectual:

> [O]bservational findings are not at all final barriers for theories, although they are usually presented as such. Observational findings can be reinterpreted, and can perhaps even be made to *lend support* to a point of view that was originally inconsistent with them. ([1965a], p. 202)

This, which Feyerabend calls 'the manufactured character of evidence', is perhaps the most radical and corrosive idea he pursued. It means we should be wary when we find him appealing to empiricism.

3.2   Feyerabend's Pragmatic Theory of Observation

The principles of pragmatic and phenomenological meaning are versions of what Feyerabend calls the 'semantic theory of observation', which characterizes observation-statements in terms of their *meaning*. It is this theory to which Feyerabend is fundamentally opposed: it is, he says 'unacceptable to anyone who rejects the synthetic *a priori*' ([1965c]: *PP1*, p. 125). In its place he proposes to put a *'pragmatic theory of observation'*, an attempt to capture what Mary Hesse calls 'the essentially causal character of observational meaning' ([1980], p. 144). Here, observation-statements are distinguished not by their meaning but by the cause of their production, by the fact that their production conforms to certain behavioural patterns.

Although he is sometimes interpreted as following Wittgenstein in this respect, Feyerabend reveals his real inspiration when he tells us that he 'started from and returned to the discussion of protocol statements in the Vienna Circle' ([1991], p. 526; AM^3, p. 212). The pragmatic theory of observation he attributes to Popper, Carnap and Neurath, who, he says, developed it in the 1930s with 'excellent', 'crystal clear' arguments.[6]

The Logical Positivists originally held the 'psychologistic' view that scientific observation-statements are about sense-data, and can be justified by perceptual experience. Otto Neurath was the first to dissent, arguing for 'physicalism', the view that observation-statements are about physical objects. Popper and Carnap soon followed him. In chapter 5 of *The Logic of Scientific Discovery*, Popper tried to exorcize the remaining elements of psychologism from this new physicalist theory of observation. According to him, the 'basic statements' of science do not refer to and are not based on perceptual experiences, but are equivalent to statements about the relative positions of macroscopic physical bodies. Statements of this kind, observes Feyerabend, are such that 'a normal observer can quickly decide whether or not an object possesses such a property' ([1962a]: *PP*1, p. 50). (They are easily decidable, in the sense of chapter 2.) But, notoriously, Popper remained ambivalent over the role and status of basic statements, and the Logical Positivists, according to Feyerabend, abandoned the pragmatic theory of observation, returning to their old psychologistic sense-datum ideology.

Feyerabend's pragmatic theory of observation, regardless of whether the positivists ever really held it, is the most uncompromising attempt to develop Popper's line of argument, and to extirpate from it what remains of the idea that experiences, or observation-statements, have an 'empirical core' of meaning that is independent of theory. For Feyerabend, observation-statements are not semantically different from other contingent statements: they have no special kind of meaning or special core of content. Meaning and observability are two different and independent dimensions of a term. The only difference between observation-statements and other statements pertains to the psychological or physiological circumstances in which they are produced. Because these circumstances are observable, we can determine whether a bodily movement is correlated with an external event and can therefore be regarded as an indicator of that event.

> [A] statement will be regarded as observational because of the *causal context* in which it is being uttered, and *not* because of what it means. According to this theory, 'this is red' is an observation sentence, because a well-conditioned individual who is prompted in the appropriate manner in front of an object that has certain physical properties will respond without hesitation with 'this is red'; and this response will occur independently of the *interpretation* he may connect with the statement . . . All we need in order to provide a theory with an observational basis are statements satisfying this pragmatic property . . . Their meaning they obtain from the theory to which they belong. ([1965a], pp. 198–9)

Feyerabend was not denying that there are such things as observation-statements: he was offering an account of what they would be like if we were to accept his pluralistic methodology and his materialist metaphysics.

Feyerabend's realism says that we reinterpret our experiences in the light of the theories we subscribe to. Even so, there is a pragmatic invariance insofar as we demand that even the most high-level theory yield those results contained in a low-level theory. Human reactions, the fact that certain statements are produced almost automatically under certain circumstances, thus constitute the core that all scientific theories have in common: 'The "unity of experience" is, thus, a practical unity, which is based on the unity of structure of human organisms' ([1960a], p. 72). Experience, Feyerabend suggests, can be explained in two ways: either in terms of its intrinsic properties, or in terms of the effects it has upon us:

> The latter consists in our having certain dispositions, such as the disposition to utter the noise 'red' or 'it stinks' and the like, which noises will of course have different meanings in different theories. Now if these theories are at all accessible to test, they ought at least agree with us to the extent that a robot which has been programmed in accordance with them will say 'red' or 'it stinks' under exactly the same circumstances under which we are inclined to utter these sounds. Hence, what is *common* to all good theories in a certain domain is that on application they are able to mimic, on the everyday level, the *behavioural pattern* of normal human beings. ([1961b], p. 83)

According to the pragmatic theory,

> we must carefully distinguish between the *causes* of the production of a certain observational sentence, or the features of the process of production, on the one side, and the *meaning* of the sentence produced in this manner on the other. More especially, a sentient being must distinguish between the fact that he possesses a sensation, or disposition to verbal behaviour, and the interpretation of the sentence being uttered in the presence of this sensation, or terminating this verbal behaviour. Now our theories, apart from being pictures of the world, are also instruments of prediction. And they are good instruments if the information they provide, taken together with information about initial conditions characterising a certain observational domain D, would enable a robot without sense organs, but with this information built into it, to react in this domain in exactly the same manner as sentient beings who, without knowledge of the theory, have been trained to find their way about D and who are able to answer, 'on the basis of observation', many questions concerning their surroundings. This is the criterion of predictive success, and it is seen not at all to involve reference to the *meanings* of the reactions carried out either by the robot or by the sentient beings . . . All it involves is *agreement of behaviour*. ([1962a]: *PP*1, p. 93)

As an account of how scientific theories are actually assessed, this seems deeply problematic. It may be that a well-conditioned individual, appropriately prompted in front of an object that has certain physical properties will say 'this is red'. But it is fantasy to think that this response 'will occur independently of the *interpretation* he may connect with the statement' ([1965a], p. 198), for this implies, falsely, that scientific observers are not concerned with the meaning of the observation-statements they produce. To the contrary, if an observer did not think he knew the meaning of his observation-statement, he would not regard it as the appropriate response. The production of observation-statements can be assessed in terms of reasons. What is more, the pragmatic theory is an excessively instrumentalist position for a scientific realist to commit himself to, since it suggests that the 'criterion of predictive success' is the only criterion we use to evaluate theories, ignoring others such as consistency, explanatory power, scope, simplicity, precision, convenience, unity, coherence, elegance and fruitfulness. Lastly, it buys into a stringently behaviourist conception of human beings, which Feyerabend elsewhere rejects. Indeed, in the same papers where he presents the pragmatic theory of observation, he rejects the behaviourist account of the meaning of observation-statements later proffered by Carnap. Carnap had argued that the meaning of basic observation-terms was fixed by virtue of their designating *observable* properties and relations.[7] Carnap, says Feyerabend,

> quite obviously presupposes that the meaning of observational terms is fixed independently of their connection with theoretical systems. If the pragmatic theory of observation were still retained by Carnap, then the interpretation of an observational statement would have to be independent of the behavioural pattern exhibited in the observational situation as well . . . We must . . . suspect that, for Carnap, incorporation of a sentence into a complicated behavioural pattern has implications for its meaning, i.e. we must suspect that he has silently dropped the pragmatic theory. This is indeed the case. ([1962a]: *PP1*, pp. 53–4)

We have already seen enough reason to think that a behaviourist criterion of what it is for a statement to be an observation-statement, according to which the meaning of an observation-statement is fixed by the way it is handled in situations involving observation, is preferable to Feyerabend's behaviourist theory of observation. Feyerabend's objection is not that such a behaviourist criterion of observability cannot be correct, but that it can (only?) be *made* correct by a methodologically undesirable scenario in which a language has had the meanings of its observation terms petrified: 'The behaviouristic criterion of observability will be satisfied by any language that has been *used for a long time*; a long history and the observational plausibility brought about by it are the best preconditions for the *petrification of meanings*' (*PP1*, p. 54; emphases added).

This is an important and neglected aspect of Feyerabend's view. Popper, as we saw, accepts that our 'methodological' decisions change our epistemic situation: he admits that we *can* have certainty and epistemic security, if we want it. But, as we also saw, he refused to extend this conventionalism to semantics. Feyerabend pushes further. He can admit the correctness of other theories of meaning (for instance, phenomenological, pragmatic theories), but will then argue that the real issue is whether this is the kind of meaning we want, or whether we want meanings to be set up in a way which, in fulfilling reasonable methodological demands, maximizes the progress of knowledge. If we have, by our own decisions and activities, made a positivistic theory of meaning true, Feyerabend tries to persuade us to reverse these decisions and change our conceptual scheme. But, against Feyerabend, it must be said that changing our conceptual scheme does not make possible something that was previously conceptually impossible: it just means that we no longer have ways of referring to the old conceptual impossibility.

In general, Feyerabend holds that our epistemological decisions actually realize, or create, different forms of activity. We are explicitly told that the pragmatic theory of observation is a proposal about the nature of observation-statements, not a description:

> The choice between the pragmatic theory and the semantic theory is of course purely a matter of convention. If we want synthetic *a priori* statements, if we want to be able to derive eternal laws from facts of observation ... then the semantic theory will most certainly be a Splendid Thing. However, if it is our intention not to except any part of our knowledge from revision, then we shall have to choose the pragmatic theory. ([1965c]: *PP*1, p. 125)

Just as, for Popper, science has no 'nature' independently of our decision to see it in one way or another, so, for Feyerabend, the nature of observation, the nature of meaning, and the relation between theory and observation are all determined by our decision to adopt particular theories about those matters. It is hard to see how this can be squared with the realism that Feyerabend simultaneously professes, for it subverts the idea that these 'theories' of observation and of meaning are about mind-independent phenomena, phenomena whose natures do not depend on our decisions. Although the assignment of meaning to the atomic elements of languages is conventional, once the assignment is made there are facts about observation-statements (such as the fact that they have meanings, or the fact that one such statement means so-and-so) which are not themselves conventional, and which it does not make sense to say can be decided.

The pragmatic theory of observation has come under fire ever since its inception. It is, as we shall see, supposed by Feyerabend to provide a way of choosing between high-level theories (even 'incommensurable' theories) on the basis of observation, in the absence of crucial experiments. Much of the critical fire concerns this contention, which will be considered in chapter 6.

3.3 Radical Conceptual Change

Feyerabend believes we ought to reject the semantic theory of observation because it contravenes what he calls the 'principle of revision', that we should never admit any dogmatic or incorrigible statement into the corpus of our knowledge. The sciences, he says, 'are the result of a decision to use only testable statements for the expression of laws and singular facts. This being the case we cannot admit any irrefutable statement, however elevated and noble its source may seem to be' ([1962a], pp. 39–40). This is not the only place in Feyerabend's early writings where he blatantly presupposes the descriptive correctness of his falsificationist methodology. As a conventionalist about methodology, he has, of course, no right to do so. Falsificationism, for Popper and Feyerabend, is a proposal about what the methodology of science should be. The case for its descriptive correctness has not been made.

Feyerabend wields the principle of revision against two very different kinds of statements which are, for empiricists, the locus of certainty. He uses it to argue, first, against the existence of incorrigible observation-statements, and second, against statements that are supposed to be true *a priori*. The semantic theory of observation says that the assimilation of observational data takes place within an existing framework of concepts. It thereby imparts a certain stability to some elements of our theoretical corpus, by presupposing that the grammatical rules governing the relevant terms are already fixed. The pragmatic theory of observation, by contrast, implies that we should never isolate any part of the corpus of our 'knowledge' (more accurately, our belief system) from revision, that in fundamental conceptual change there should be no inferential wallflowers which must remain unrevised.

Feyerabend argues that we should not want semantic stability, and that radical conceptual change is always preferable to minor revision. Because science only progresses via conjecture and refutation, we should relish the opportunity to test, and thus perhaps to refute, even the most fundamental members of our conceptual framework. Indeed, we must take this opportunity unless we are prepared to grant a *de facto* privileged *a priori* status to these truths, and as empiricists we must find such a course of action unacceptable. Empiricists are thus forced to choose between a foundation of incorrigible statements, governed by a framework of untouchable truths, and good scientific methodology.

Feyerabend believes that conservatism in theoretical transitions is generally indefensible. It is important to see how far this goes against most people's intuitions in this matter. Most philosophers of science accept some principle of conceptual conservatism, some general injunction to react to an anomaly facing one's theory by embracing the nearest theory unaffected by the anomaly, making the least possible revision in our conceptual scheme.[8] Feyerabend's attitude is totally and unreservedly opposed to this, and sets him at the furthest remove from most other philosophers of science. In his most Popperian work, he deems the making of minimal adjustments to be an *ad hoc* stratagem, designed to protect a theory from falsification. For this 'methodological' reason, Feyerabend advocates maximal shifts in conceptual frameworks. This radicalism about conceptual change, which runs right through his work, and which he inherits from Popper, is at variance with both the lay person's feeling that scientific theories are of enduring value and with the usual conceptually conservative demand that new theories should be minimal revisions of old ones.[9] It is engendered by faith in the notion of scientific revolutions. Feyerabend borrowed this concept from the new historians of science. The term was originally applied to a single event, 'the' scientific revolution (located by Koyré in the seventeenth century, by Butterfield within the period from 1300 to 1800). Kuhn explicitly broadened the concept by applying it to different events in the history of science. For him and Feyerabend, a scientific revolution is an event in which old patterns of thought are totally overthrown, and intellectual activity is rebuilt from its foundations. For these philosophers, such events are the most important ones in the history of thought. Feyerabend's fundamental message is that we have the choice of seeing our theoretical inheritance as a stable edifice which has been added to only peripherally by the advent of new theories, or as a succession of theories which we have outgrown. Neither vision is correct or incorrect to pre-existing phenomena, but the former advances science more than the latter, and thereby plays up to our *ideals* of inquiry. However, Feyerabend and Kuhn go beyond Popper in demanding not just that successive theories should be genuinely new conjectures, but also that they should sometimes take root within new conceptual frameworks, and thus lack any *semantic* continuity with their predecessors. In chapter 6 we shall see the problems that result.

3.4   Humans as Measuring Instruments

Feyerabend sometimes introduced the pragmatic theory of observation by claiming that it is trivially true of measuring instruments. The readings of such instruments do not 'mean' anything unless their users have a *theory* about their operation which posits a reliable correlation between certain physical situations and the indications of the instrument. That theory supplies the interpretation of the instrument's readings, and a large enough change in the theory will entail a change in our interpretation of the readings.

A pragmatic theory of observation may be appropriate for measuring instruments (and may, incidentally, be an illuminating way of looking at subjects like 'computer vision'). But Feyerabend wants to give a purely causal theory of *human* perception, modelled on this causal theory of machine 'perception'. He argues that the only reason for resisting such a theory comes from classical foundationalism, which he characterizes as

> the (very old) belief that (a) some states of the mind (sensations or abstract ideas) can be known with certainty; that (b) it is exactly this knowledge that constitutes the foundation of whatever assertion we make about the world; and that (c) meaning invariance [obtains]. ([1962a], p. 38)

Such ideas, he suggests, can be eliminated when we realize that classical foundationalism obliterates the crucial distinction between facts and conventions. Conversely, the attempt to uphold this distinction leads to the separation, characteristic of the pragmatic theory, of the observational character of a statement from its meaning. The pragmatic theory is thus supposed to be a consequence of the distinction between nature and convention.

At this point we must protest, on behalf of the semantic theory of observation, that a statement's epistemological status is partly constitutive of its meaning. An observation-statement cannot change its epistemological status (its being an observation-statement) while retaining the same meaning. If one understands a statement, having some clear conception of observation-statements in mind, one thereby knows whether or not it is an observation-statement. If one did not know whether observation was relevant to deciding a statement's truth-value, one could not be said to know the meaning of that statement, one would be unable to give an acceptable explanation of its meaning, and one's use of that statement would be revealed as non-standard (in at least some contexts). More generally, what Feyerabend misses is that there must be a framework of rules underlying language if meaning is to be possible.[10] It is ironic that one who insists so strongly on the distinction between nature and convention, facts and rules, should fail to see that language must comprise rules which do not amount to factual claims, in order that the making of factual claims should be possible.

The pragmatic theory of observation implies that a theory can be tested by assessing its ability to mimic the linguistic and behavioural patterns of normal human beings, who are to be regarded, in this respect, as measuring instruments. Feyerabend, who may have got this idea from Goethe, set the ball rolling thus:

> What the observational situation determines (causally) is the acceptance or the rejection of a sentence, i.e. a physical event. In so far as this causal chain involves our own organism we are on a par with physical instruments. ([1958a]: *PP*1, p. 19)

> The living brain is already connected with a most sensitive instrument – the living human organism. Observations of the reactions of this organism, introspection included, may therefore be much more reliable sources of information concerning the living brain than any other 'more direct' method. ([1963b]: *PP*1, p. 166)

> As opposed to many alternative accounts, the pragmatic theory of observation takes seriously the fact that human beings, apart from being called upon to invent theories and to think, are also used as measuring instruments. ([1965a], p. 212)

Feyerabend explicitly links the pragmatic theory of observation both to the project of scientific materialism and to scientific realism. 'Science', he says, 'has long taken it for granted that the human body, the human mind, and perhaps all of man can be explained on the basis of materialistic principles' ([1965a], p. 213). The pragmatic theory, he goes on,

> restores to science the right to examine human beings according to its own ideas. Moreover, it assumes that the *interpretation* of the observation sentences is determined by the accepted body of theory. This second assumption removes the arbitrary barriers and the *a priori* elements characteristic of the idea of the observational core. It encourages us to base our interpretations upon the best theory available.

Recently, Paul Churchland has picked up and extended this theme in a very thorough way.[11]

A proponent of the rival semantic theory of observation, however, might well object that although our internal life is regular, and we react regularly to certain stimuli, humans are not rightly conceived of as measuring instruments. To think of ourselves so is to conflate the nomic regularity of causation with the normative regularity of rule-governedness. The pragmatic theory conflates what H. P. Grice called 'natural' and 'non-natural' meaning,[12] thereby subverting the latter, which is the concept appropriate to linguistic meaning. It involves a denial of the normative, rule-governed aspect of language.

Why should we think the pragmatic theory can be extended to cover human beings? A vital difference between human perception and animal or robot perception is that whereas in the latter cases perception can be described in purely causal terms, in the former case the presence of a perceivable X gives a person a reason to believe that an X is present. This is why scientific observation-statements are regarded and assessed as responses appropriate to what the observer takes himself or herself to be observing, not just as reactions to stimuli. The fact that it takes place within the 'space of reasons' makes perception by fully socialized humans a qualitatively different matter from the comparatively low-grade registration of information by measuring-instruments, photoelectric cells, non-language-users, etc. The pragmatic theory of observation pointedly ignores this extra dimension to human perception.

3.5 The 'Problem of Theoretical Entities'

Feyerabend seems to think positivism at its strongest when based on a sense-datum epistemology, but he claims to have shown, in his 1960 paper 'The Problem of Theoretical Entities',[13] that a sense-datum epistemology is untenable. Here he enquires after the difference between observational and theoretical concepts, and takes sense-datum epistemologies to yield the strongest defence of the stability-thesis. Sense-datum theories are, of course, paradigm examples of the semantic theory of observation.

The 'problem' of theoretical entities is whether there are objects which correspond to theoretical concepts. Can we say that, like certain concepts of observable things, the role of theoretical concepts is to *refer to* objects? Must we suppose that 'theoretical entities', such as electrons, quarks, genes and the unconscious really exist? Must we suppose that if our theory, which contains both observational and theoretical concepts, is true, then its theoretical terms refer? Positivists took this problem very seriously, but Feyerabend held that the alleged problem of theoretical entities is unstable, that it collapses in different directions depending on how we define 'observational' and 'theoretical'.

He argues that these questions about the status of theoretical entities only make sense if we have already made two assumptions: (a) that the existence of theoretical entities is problematic only because they cannot be observed, and (b) that there *are* unobservable theoretical entities. These problems will be solved, or at least clarified, not by scientific investigation, he thinks, but by exhibiting 'the *value* of theoretical knowledge or . . . scientific methodology' ([1960a], pp. 36–7).

Three explanations of the concept of an observation sentence are considered. The first says that the concept of an X is observational if its associated singular sentence ('That's an X') is easily decidable on the basis of observation:

> A concept is an observational one if the truth-value of a singular sentence containing either only that concept or that concept along with other observational concepts can be arrived at quickly and solely on the basis of observation, or at least if it's possible to imagine that a decision of this type will one day be possible. ([1960a], p. 35)

Are there theoretical entities in this sense, entities which are forever incapable of being observed? Feyerabend argues that there are no such things, that some theoretical terms denote what for *some* observers are observable entities, but that, for example, cloud chambers actually allow us to observe elementary particles. Although it is often maintained that one cannot see an electron (since electrons are too small in relation to the wavelength of light), new rules for this use of the term 'see' can, he says, be explained to everyone. He even goes so far as to say that 'we can demonstrate that adherence to the principles of scientific method must in the end lead to the direct observability of the states of affairs asserted by the theory. Empirical method does demand, after all, that each statement of a physical theory must be made available for verification by experience' ([1960a], p. 38). Although this sounds uncharacteristically verificationist, and doubly odd coming from one who believes that science has no determinate 'empirical method' of its own, what Feyerabend means here, I think, is only that our theories, in conjunction with ever-improving experimental apparatus, allow us to make decisions about the presence of theoretical entities which were once unobservable. We can come to adopt an observational criterion for the presence of an initially unobservable entity. The result is supposed to be that in this sense of the term 'observable', all theoretical entities are ('in principle') observable. This result refutes assumption (b), that there are entities forever incapable of being observed. Thus this first version of the 'problem' of theoretical entities collapses.

Feyerabend considers objections to this position, the most powerful being that while theoretical entities are knowable by inference, they are not directly observable. We only come to know of the presence of an electron in a cloud chamber by mobilizing a sophisticated theory of what is going on in such a device. Feyerabend replies that the distinction between knowledge by direct observation and knowledge by inference is merely a distinction between different stages of the subject's learning. The processes of conscious inference which are initially necessary to come to the conclusion 'Lo, an electron', are gradually washed away by familiarity with the apparatus, to be replaced by non-inferential or direct observation.

Such an answer, however, presupposes that the issue is a genetic or psychological one, and defenders of the distinction between direct and indirect observation certainly would not concede this. They would insist that the issue is over the justification of the putative observation-statement: only if its justification proceeds without reference to theory are we dealing with a genuine observation-statement. If the statement can only be justified by reference to theory, the existence of the theoretical entity mentioned is not secured, since the truth of a theory can never be guaranteed but is always 'hypothetical'.

Feyerabend agrees that this is the most important objection, but urges that it involves a *change* in the concept of an observation-statement proffered by the first explanation. It involves thinking of observation-statements as certain, as opposed to hypothetical. So the second explanation of the concept of an observation-statement he considers is that

> a concept is observational when a singular sentence which contains it (or the concept on its own) is entirely derived directly and without any thought. It must also require no further *justification* than that obtained by pointing out that a certain observation has been carried out. Observation sentences are thus certain and not hypothetical. ([1960a], p. 41)

Feyerabend now argues that this second form of the problem of theoretical entities also collapses in upon itself, since assumption (a), that the existence of observable entities is not a problem and that the existence of theoretical entities is only a problem because they are not observable, is no longer met: all concepts are theoretical concepts in the sense of the second explanation. The apparently humble concept of a table, for example, is a theoretical concept because our ability to perceive tables 'depends on our having learnt the efficient use of a very complex instrument – the eye' ([1960a], p. 42). What is more, our perception of tables 'depends on the nature of the intervening medium and also upon the laws of the propagation of light within this medium [. . . and also upon] the momentary psychological state of the observer' (p. 42). All these factors play a part in the justification of perceptual propositions about tables. If we decide to use the pragmatic theory of observation, we will not admit into our language any observation sentences in the sense of the second explanation. So the theory of sense-data cannot save the problem of theoretical entities in its second form from absurdity.

These considerations do not establish the point at issue. That point is whether use of our eyes relies on our deploying a theory. The fact that perception only takes place under certain conditions does not mean that our concept of what is perceived includes a specification of those conditions. We do not have to have theoretical knowledge of, for example, the eye's functioning in order to be able to see tables. On the contrary: while all of us are capable of perceiving tables in one way or another, very few of us have theoretical information about the physiological and psychological processes subserving perception. So whether or not the factors Feyerabend mentions can play a part in the justification of perceptual propositions, the fact that none of them are theories (properly so-called) means that 'I see a table' is not a theoretical statement, and that the concept of a table is not a theoretical concept. Feyerabend has not shown that this variant of the problem of theoretical entities collapses, or that 'the problem of the existence of theoretical entities stands and falls with the correctness of the theory of sense-data' ([1960a], p. 42).

In fact, the whole issue of certainty, which Feyerabend concentrates on, is a red herring. We do not have to add the dubious (and characteristically empiricist) rider that theories are hypothetical and observation-statements are 'certain'. The objection to Feyerabend's position simply relies on there being a distinction between theories and other kinds of statements. As long as there is such a distinction, the distinction between theoretical statements and observation-statements is available. The first explanation of the concept of an observation-statement has not been invalidated.

3.6 The Critique of Sense-Datum Epistemologies

Feyerabend put forward another critique of the idea, which we might call the *indubitability thesis*, that there are empirical sentences whose truth cannot, under certain circumstances, be doubted. Traditionally empiricists have, in espousing sense-datum epistemologies, granted a privileged role to statements about one's own sensations and perceptual experiences. Aiming to demonstrate that even such statements are subject to doubt, Feyerabend starts by remarking that not *all* statements about feelings exclude doubt:

> Whoever has undergone a sense-test after partial paralysis will know how difficult it is to distinguish between the feeling of being touched by a sharp object and a blunt object. Note that what is in doubt is not only the conclusion concerning the object but also the correct recognition of the sensation itself. It is sometimes difficult to decide whether the feeling was painful. ([1960a], p. 44)

The reason he gives for this is that sensations are not 'absolute', they are *contrast*-phenomena. They appear against a background or 'sea' of other feelings which, because they do not enter consciousness, require special attention for their analysis. This, he claims, applies also to sense experiences.

Feyerabend recognizes that these examples may be exceptional, since the sensations involved are feeble. But he argues that we can mis-identify even the most intense sensations. This is what we should expect if sensations are contrast-phenomena: as soon as the background exceeds a certain intensity, identification of the foreground phenomenon becomes problematic. As an example, we are asked to consider standing on an airfield close to a jet engine. The noise, he says, will be very disagreeable and even painful – but there is a point at which it is unclear whether it is noise or pain that is being experienced. The obvious response here is 'Why not say that it's *both*?' Or rather that the noise is so loud that it (the noise itself) is painful? Because pain and noise are not determinates of the same determinable (pain is a sensation, noise is not), their joint presence is not ruled out.

This is the correct response also to the second example Feyerabend mentions, Bishop Berkeley's case of a feeling of heat so intense that it merges into a feeling of very great pain.[14] Feyerabend contends that there is a point at which it is impossible to decide whether what is being experienced is a feeling of heat or of pain. But both Berkeley's own contention, that the heat and the pain have become one, as well as the more correct conclusion that the heat has become painful, show that, *pace* Feyerabend, this *isn't* an example where one is in doubt about whether it is heat or pain one is experiencing.

As a third example, Feyerabend considers the masochist, for whom pain gives rise to pleasure. Here, he contends, it is no longer possible to establish whether intense pain or pleasure has occurred. But the whole point about masochists is that they get pleasure *from* pain, so that the ordinary contrast between the two kinds of sensations is blurred.

In 1960 Feyerabend also argued against the sense-datum theory in two other ways. First, if we decide to see humans as measuring instruments we will not admit any 'certain' or 'incorrigible' statements into our language, and thus there will be no sense-datum statements as traditionally conceived, no statements which obey the stability-thesis.

Second, Feyerabend argued that if we were to take the wrong decision here we would have a sense-datum language, but that it could not comfortably play the role of an intersubjective observation-language for science. Observation-statements, as conceived here, cannot be made public or exposed to testing, they 'are mostly meaningless and are meaningful only in isolated moments and even then only for some few individuals who, however, can never communicate with each other' ([1960a], p. 60). He therefore urged us not to use such a language, for its use would amount to the elimination of science.[15] He summed up these two arguments by saying that 'the theory of sense-data cannot save the problem of theoretical entities, in its second form, from absurdity, because *methodological considerations demand the elimination of sense-data*' ([1960a], p. 61; my emphasis). By 1961, he was convinced that we must disregard sense-data, as their non-existence had 'definitely been established' ([1961c], p. 82).

The third and final explanation of the concept of an observation-sentence that Feyerabend considers leaves the sense-datum theory behind, and simply identifies observation-sentences as sentences which obey the stability-thesis. This is an attempt to find a concept of an observation-sentence which is intuitively obvious, corresponds to scientific practice, and is not such that every concept is either observational (as is allegedly the case in the first explanation) or theoretical (as in the second). It was defended by Feigl, who argued that because observation-sentences are used to decide between rival scientific theories, they must themselves be stable in meaning and independent of theory. Feigl proposed that ordinary language, being relatively free from theoretical penetration, satisfies the stability-thesis, and therefore makes a suitable scientific observation-language (see *PP*1, pp. 85–6, 152 n. 18). Feyerabend disagreed. He correctly pointed out that there is a gap in Feigl's argument, since it does not follow that in order to test theories the meaning of observation-sentences must be independent of *any* theory, but only that it must be independent of the theories that are being compared. However, this suggestion does not solve the problem about high-level theory comparison, since high-level theories are precisely the ones which, on Feyerabend's view, *will not* disagree over synonymous observation-sentences. Not even relatively theory-neutral observation-statements are available in 'crucial experiments' between high-level theories, since such theories will, according to Feyerabend, determine the meaning of even the most low-level observation-statements.

Feyerabend also retorted that ordinary language, although relatively stable, is not an observation-language in Feigl's sense, since it does contain substantial theoretical elements But his argument for this is weak, since he thinks that we have shown the theoretical status of ordinary language if we can show that a change in that language is imaginable. On the contrary: no one needs to defend the idea that everyday language is in principle unchangeable. The most that a defender of a reasonable stability-thesis needs is that radical conceptual changes at this level of language are not necessitated by higher-level theoretical changes. Feyerabend concludes this argument by saying that anyone who accepts the principle of revision has to reject the stability-thesis and the existence of an observation-language (in the sense of this third explanation). Problems in the theory of perception, he says, are 'not solved by *proofs*, but by *decisions*' ([1960a], p. 65).

The strongly normative flavour to all these arguments is apparent, and so, perhaps, is their potential weakness. Methodological criticisms are all very well, but if one appeals to scientific methodology, one has to have got the methodology correct in the first place. One cannot defeat metaphysical views by complaining that they fail to measure up to methodological rules which are not the rules scientists accept. Because Popper and Feyerabend do not show the descriptive correctness of their proposed methodology, their criticisms of foundationalism remain largely unsupported. The only one which retains its force is the very reasonable complaint that a sense-datum language would be an unsuitable scientific observation-language.

4

Scientific Realism and Instrumentalism

4.1 Feyerabend's Scientific Realism

Feyerabend's unorthodox version of scientific realism, which remains relatively constant through his early writings, comprises the following ingredients. First, the idea that our theories do not stop at experience or at appearances (as positivists might have it), but go beyond, to explain these things in a deeper way: 'According to the realistic interpretation, a scientific theory aims at a description of states of affairs, or properties of physical systems, which transcends experience . . .' ([1960b]: *PP1*, p. 42). In Feyerabend this becomes what Popperians call 'conjectural realism', the idea that our theories are attempted descriptions of the world, or of reality, descriptions which can be evaluated as true or false. This represents a commitment to the existential character of scientific theories. Feyerabend even goes further, insisting that theories tell us about what things *are*, their very *nature*, in a world which exists independently of measurement and observation. All these ideas comprise what we might call the ontological ingredient in his scientific realism.[1] But although Feyerabend endorsed conjectural realism, he never really embraced the important Popperian claim that our theories exhibit increasing correspondence with reality ('verisimilitude', or convergence to the truth). This makes Feyerabend's realism immune to attacks on richer forms of that doctrine.[2]

Second, Feyerabend's realism has a semantic ingredient, according to which theories are universal statements used for explaining facts, statements whose terms have what he calls 'direct factual reference', or 'descriptive meaning'. Realists propose to take theories literally, without extensive reinterpretation. Although Feyerabend recognizes that theories *can* be given differing interpretations, he argues for what he calls 'realistic' interpretations, interpretations in which the theory is understood on its own terms: '*The interpretation of a scientific theory*' he says, '*depends upon nothing but the state of affairs it describes*' ([1960b], *PP*1, p. 42).[3]

Lastly, there is a strong epistemological or psychological ingredient whose nature is harder to pin down. It is encapsulated in 'Thesis I', which, as stated so far, involves Feyerabend's semi-technical term, 'interpretation'. But later statements of this idea, which is his alternative to the stability-thesis, do not depend on the machinery of the earlier paper. For Feyerabend, it is *observation-sentences* that are in need of interpretation and *not* theories:

> [T]he meaning of observation sentences is determined by the theories with which they are connected. Theories are meaningful independent of observations; observational statements are not meaningful unless they have been connected with theories. ([1965a], p. 213; see also [1960d], p. 247)

In the realist tradition as Feyerabend understands it, facts of experience are not treated as unalterable building-blocks of knowledge but as capable of, and crying out for, analysis and improvement. The 'observational level', or the level of 'ordinary language', is part of the theoretical level, rather than something self-contained. A realist, says Feyerabend,

> wants to give a unified account, both of observable and of unobservable matters, and he will use the most abstract terms of whatever theory he is contemplating for that purpose . . . He will use such terms in order to either *give* meaning to observation sentences, or else to *replace* their customary interpretation. ([1970a]: *PP*2, pp. 153, 155; *AM*[1] p. 279)

This demand for a *unified* account of disparate domains is important. Feyerabend seems to think that the realist, aiming to describe the nature of things, must strive towards what is now known as a unified 'theory of everything'.

Consider, for example, the phenomenon Feyerabend calls the '*duality*' of light and matter'. Some experiments lend support to a particle theory of light, and point away from a theory according to which light is wavelike. Others point in exactly the opposite direction. A positivist can interpret this duality as something 'given', and not in need of explanation. He will not find it necessary to seek a new and unified theory, because for him explanation is incorporation into a predictive scheme, and the two available schemes are, when confined to their respective experiments, predictively successful. But duality is a serious challenge for a realist who wants to think of light (and matter) as 'something which is fundamentally a single and objective entity' ([1958b], p. 78); she must consider existing theories inadequate, and search for a new one which is (a) empirically adequate and (b) *universal*, that is, 'it must be of a form which allows us to say what light *is*, rather than what light appears to be under various conditions'. For the Feyerabendian realist, the solution to the problem of duality is therefore found in the attempt to devise a completely new universal theory.

4.2 Instrumentalism

Feyerabend sees the positivist/realist dispute in the traditional way, as a disagreement over the cognitive status of theories. His argument for scientific realism, in its broadest contours, is as follows: scientific progress is desirable; it is best furthered by theoretical proliferation; and scientific realism leads to a proliferation of theories, while positivism does not. This is clearly a development of Popper's suggestion that positivism is *heuristically infertile*, that it would produce bad science. Feyerabend accepted this suggestion, but considered it inadequate since it is merely a generalization from the past history of science. Popper may have been correct to say that realism has encouraged progress, but the inductively-based conjecture that it will continue to do so in future is too weak to justify realism (not to mention the fact that reliance upon such an inductive argument is hardly in keeping with the tenor of Popper's philosophy). Instead, Feyerabend aimed to show that realism must always be a more heuristically fertile strategy of theory interpretation than its main positivist rival, *instrumentalism*, the view that 'scientific theories are instruments of prediction which do not possess any descriptive meaning' ([1958a], *PP*1, p. 17), or that we should stop interpreting and regard theoretical statements as 'cognitively meaningless instruments of prediction' ([1958c]: *PP*1, p. 240).[4] Feyerabend characterizes instrumentalism as Popper did:

> Instrumentalism maintains that [a] theory must not be interpreted as a series of statements, but that it is rather to be understood as a predictive machine whose elements are tools rather than statements and therefore cannot be incompatible with any principle already in existence. ([1962a]: *PP*1, p. 83)

Jerzy Giedymin complained that this portrays it as 'up to the scientist (or philosopher) to decide the status of theoretical statements' ([1976], p. 201). Feyerabend would cheerfully admit this charge, since he applies his conventionalism across the board. (Recall his insistence that we can decide the form of our theoretical knowledge.) But Popper is not in the same boat, since he is only a conventionalist about methodological issues. Feyerabend's innovation is to extend this conventionalism to the issue of the interpretation of scientific theory.[5]

However, the question of the interpretation of a scientific theory or law is not, *pace* Feyerabend, a matter of convention. Questions about the interpretation of science are resolvable by a correct understanding of scientific practice. Whether theories are descriptions or mere instruments is a conceptual (and therefore factual) issue, and calls for an investigation into their actual functions, not for a decision. Like normative epistemology, conventionalism about the realism/ instrumentalism issue is ultimately based on something like rule-scepticism: the idea that there is 'no fact of the matter' as to which rule a person is following. And rule-scepticism is self-defeating.[6] The fact of the matter as to whether theories, laws and statements are attempted descriptions of reality or mere instruments of prediction is discovered by finding how they are *used*. (Of course, we have to allow that philosophical views can influence scientific activity: if scientists read enough positivist or realist philosophy, their reasoning and practice may change as a result. So philosophical views are not *just* 'higher-order'. But even if we asked scientists to treat theories and laws in a different way in future, and they did so, this would not establish their present nature.)

At least Feyerabend has the virtue of consistency here. Popper officially refuses to apply methodological conventionalism to the positivist/realist dispute, holding that this dispute can be solved (conceptually). But in fact his objections to positivism are purely methodological: he treats the issue of realism vs. positivism as one concerning which is the better scientific methodology.

As we noted in the previous chapter, Feyerabend thinks positivism at its strongest when based on a sense-datum epistemology. This is not incompatible with his assumption that instrumentalism is the dominant positivist view of the cognitive status of theories, since he also holds the connecting thesis that to accept a universal (or 'global') instrumentalism is to accept the sense-datum account of knowledge characteristic of foundationalism ([1962a]: *PP1*, pp. 83, 120). Because of this, his case against instrumentalism rests partly on the arguments against sense-datum epistemology already discussed.

But this connecting thesis is false, for while it is difficult to motivate instrumentalism without some kind of commitment to verificationism, there is no essential connection between instrumentalism and a sense-datum epistemology. One can, for example, hold the empiricist thesis that human knowledge only extends to the observable parts of the world, without holding the *radical* empiricist thesis that it extends only to our sense-data.[7] In fact, Feyerabend's connecting thesis is falsified by something he himself also accepts, that the motivation behind global instrumentalism is that 'only observation terms are candidates for a realistic interpretation' ([1964a]: *PP1*, pp. 185–6). Feyerabend's connecting thesis would work only if observation terms refer to sense-data, not, for example, to physical objects. But we have already agreed with him that the observation language of science is not a sense-datum language.

Feyerabend accepts Popper's identification of the philosophical assumption behind instrumentalism, that only observation terms are candidates for a realistic interpretation. He also urges that Popper has refuted this assumption once and for all.[8] In early papers he complains that to deny that situations described with theoretical terms exist or can be regarded as causes 'completely disregards the existential character of general scientific theories' ([1960b]: *PP1*, p. 38), and he reiterates Popper's methodological objection that global instrumentalism is an *ad hoc* manoeuvre designed to protect an ailing theory from refutation ([1962a]: *PP1*, p. 76). Thus his case against instrumentalism rests *partly* upon Popper's.

But for Feyerabend this is not the end of the matter. Although he considers Popper to have definitively refuted the idea that instrumentalism is preferable to realism,[9] he does not believe that instrumentalists must rely on the bad arguments which Popper foists on them. He therefore seeks to defend certain alleged instrumentalists, such as Bellarmino, Osiander and the founder of quantum mechanics, Niels Bohr, against Popper's criticisms, chiding Popper for neglecting their arguments in his own critique of instrumentalism:

'The view of physical science founded by Cardinal Bellarmino and Bishop Berkeley has won the battle without a further shot being fired' writes Popper. This is simply not true. First of all there is a tremendous difference between the instrumentalism of Bellarmino and the instrumentalism of Berkeley. The instrumentalism of Bellarmino *could* have been supported by physical arguments drawn from contemporary physical theory. The instrumentalism of Berkeley could not have been so supported and was of a purely philosophical nature. ([1962b], p. 261 n. 49. See also *PP1*, pp. 190, 279)

(Feyerabend does, however, credit John Watkins with the realization that although philosophical arguments for realism are insufficient, they are not therefore unnecessary.)

According to Feyerabend, the most powerful arguments for instrumentalism are certain 'physical' (factual or scientific) arguments which scientists use to show that a particular theory cannot admit of a realistic interpretation. Such an instrumentalism, although it is only a local (or 'restricted') and not a global instrumentalism, is a genuine rival to realism, not just a 'preferred mode of speech'.[10] In 'Problems of Microphysics' (1962) and 'Realism and Instrumentalism' (1964), he takes two examples of theories whose instrumentalist construal has been defended in this way: Copernicus' heliocentric theory, and quantum mechanics.

4.3   Astronomical Instrumentalism

Feyerabend stresses that Aristotelian dynamics, which heliocentrism had to overcome, was a successful and well confirmed scientific theory. It had proven problem-solving power, gave results confirmed by our everyday experience and, generally, had to its credit empirical success, theoretical success, comprehensiveness and detail. These are, as he says, weighty arguments in favour of the Aristotelian point of view. Of course, the theory had its problems, but Feyerabend, who was at this time becoming a more sophisticated falsificationist, pointed out that all theories have their problems:

> [T]heories . . . are successful in a number of cases and will be regarded as revolutionary if these cases have been troublesome for a considerable time. But there always exist *other* cases which are *prima facie* refuting instances of the theory but which are put aside, for the time being, in the hope that a favourable solution will be forthcoming. Now if we postulate that a theory which is problematic because of the existence of *prima facie* refuting instances must not be used in cosmological arguments regarding the existence or non-existence of certain situations, then we shall thereby eliminate not only the Aristotelian point of view *but every succeeding physical theory as well*. ([1964a]: *PP*1, p. 178 n. 4)

Aristotelian dynamics, however, was incompatible with the heliocentric theory; in particular, it was incompatible with the idea that the earth moves. According to the Aristotelian law of inertia, 'Everything that is moved is moved by something else. Every motion needs a mover, and this mover must be present in the close neighbourhood of the changing thing, as action at a distance is impossible. Conversely, the state of an object that is not under the influence of forces is a state of rest' (PP1, p. 177). If this is so, it is easy to show that the earth cannot rotate, since only things in contact with it would be carried along by its motion. Anything disconnected (birds, clouds etc.) would immediately assume its natural state of motion (namely, rest) and would be left behind. Since this manifestly does not happen, we must conclude that the earth does not move. This dynamical argument against the motion of the earth, which Ptolemy deployed, seems impeccable.

What, then, did the Copernican hypothesis have going for it at the time? Very little, according to Feyerabend. Admittedly, it supplied a simpler explanation of some features of planetary motion than the geocentric picture, but it did not on that account alone lead to better predictions. It only yielded empirically adequate predictions when conjoined with details which had little rationale within the heliocentric scheme. 'There was no independent evidence in favour of the heliocentric theory; this theory was, at least initially, a conjecture that had no foundation in empirical fact. The only favourable remark that could be made was that it somewhat simplified calculations' (PP1, pp. 182–3). Given that its *only* merit was purely instrumental, the idea of interpreting it instrumentalistically was unimpeachable. The dynamical arguments apparently show that the heliocentric hypothesis cannot be *true*, but can at most be an instrument of prediction. They amount, says Feyerabend, 'to a straightforward *refutation* of the Copernican hypothesis. If we take these arguments at face value, then we must regard this hypothesis as false' (PP1, p. 183 n. 15). If, as realists, we protest that the Copernican hypothesis is a true description of the actual situation, we must realize that our protest amounts to upholding an unsupported conjecture in the face of fact and well-supported theory.

4.4 Quantum Instrumentalism

In early papers, Feyerabend argued that quantum instrumentalism, the view that quantum mechanics is a tool for producing predictions rather than a theory for describing the world, implies both that the classical and the quantum 'levels' are distinct, and that the transition from one to the other is unanalysable. But he also argued that this implication is false, that the classical level is part of the quantum level, and therefore that quantum instrumentalism must give way to a realistic interpretation of the quantum-mechanical formalism. If it is replied that quantum mechanics just *is* an instrument of prediction, rather than a description ([1958b], pp. 88–9), Feyerabend is ready to concede that this might be true, thus retracting the Popperian objection. He says that quantum instrumentalism is *'not a philosophical manoeuvre that has been wilfully superimposed upon a theory which would have looked much better when interpreted in a realistic fashion. It is a demand for theory construction that was imposed from the very beginning and in accordance with which, part of the quantum theory was actually obtained'* ([1962b], p. 265 n. 62). But he counters that since theories which do admit of a realistic interpretation are definitely preferable to those which do not, the quantum theory itself must be replaced. The Feyerabendian realist is no mere philosopher; he

> cannot rest content with the general remark that theories just *are* descriptions and not merely instruments. He must then also revise the accepted *physics* in such a manner that the inconsistency is removed; i.e. he must actively contribute to the *development* of factual knowledge rather than make comments, in a 'preferred mode of speech', about the *results* of this development. ([1964a]: *PP1*, p. 177)

In a salutary critique of the pretensions of science to substantiate conclusions that cannot be established empirically, Feyerabend resisted the idea that quantum mechanics, in its Copenhagen Interpretation, can be proven to be the only acceptable theory of the microworld, arguing that while complementarity is acceptable as a heuristic picture it is pernicious if wielded as a philosophical principle. But he went on to *defend* the Copenhagen Interpretation against Popperian accusations that it was merely the result of bad (positivist) philosophizing, and that it was *ad hoc*. He sought to show that beneath a veneer of unacceptable positivist dogma there were good scientific reasons why the quantum theory was not susceptible of a realist interpretation.

One reason was as follows. Consider an interaction between two mechanical systems, A and B, in which energy is transferred from A to B. Bohr and Sommerfeld's 'old' quantum theory suggested that this transfer takes a finite amount of time, that A and B change their state, the former gradually losing energy in moving from state 2 to state 1, and the latter gradually acquiring energy in moving from state 1 to state 2. However, such a mode of description, Feyerabend says, 'is incompatible with the *quantum postulate* according to which a mechanical system can only be in either state 1 or state 2 ... and is incapable of being in an intermediate state' ([1964a]: *PP1*, p. 187). Bohr, he continues, resolved the difficulty by stipulating that during their interaction 'the dynamical states of both A and B cease to be well defined so that it becomes *meaningless* (rather than *false*) to ascribe a definite energy to either of them'. Feyerabend insists, against Popper and others, that Bohr's solution is a physical hypothesis, rather than a positivist scruple about meaning, knowledge, observability or predictability. What it excludes is not mere knowledge of intermediate states, but their very existence. (He points to other examples of concepts which can only be meaningfully applied if certain conditions are first satisfied, e.g. scratchability.) But if we think about the motion of A, or of B, we see that it too can no longer be well defined: it is no longer possible to ascribe a definite trajectory to elements of A or B. So Bohr's solution, which is empirically adequate, implies 'an instrumentalist interpretation for any future quantum theory that works with state descriptions which are well defined' (*PP*1, pp. 189–90). Apparently, therefore, 'within the quantum theory the instrumentalist position has been *forced* on the physicist by the realisation that the current theory interpreted realistically *must lead to wrong results*' (p. 190 n. 32; emphases added). Just as with the case of heliocentrism, then, '[A]ny attempt to give a realistic account of the behaviour of the elementary particles is bound to be inconsistent with highly confirmed theories. Any such attempt therefore amounts to introducing unsupported conjectures in the face of fact and well-supported physical laws' (p. 195). Feyerabend deems this a very forceful argument indeed in favour of instrumentalism.

4.5 Musgrave on Feyerabend's Defence of Instrumentalism

Alan Musgrave has pointed out that, unfortunately for Feyerabend, the idea of supporting instrumentalism with such 'physical arguments' makes no sense.[11] Because they show at most that a particular theory should not be given a realist interpretation, they can only be arguments for a local instrumentalism, instrumentalism about a particular theory. But such arguments actually presuppose realism as a global thesis! They presuppose that the laws, theories and observations which our theory conflicts with are interpreted as true statements, not merely instruments of prediction. If it were not so, there could be no clash between our new theory and the old one, there would be no motivation for refusing to construe the new theory realistically. (For example, it was because Aristotle's physics was interpreted realistically and accepted as true that Copernican astronomy had to be given an instrumentalist interpretation.) Local instrumentalism is therefore fully compatible with a general predisposition to realism. It can be the response of a realist who is disappointed to find that a particular theory cannot be a true description of reality. But such disappointed realists are not (genuine, global) instrumentalists. So when Feyerabend goes beyond Popper in this respect, he fails to show that instrumentalism should be taken seriously as a philosophical thesis.

4.6 Feyerabend's Attack on Instrumentalism

Fortunately, this does not matter for the broad sweep of Feyerabend's argument, since his defence of quantum instrumentalism, and indeed of the general possibility of local instrumentalism, was only tactical. He defended the instrumentalist's verdict that in heliocentric astronomy and quantum mechanics the attempt to impose a realist construal amounts to upholding an unsupported conjecture in the face of fact and well-supported theory. But his defence was a limited and partial one, for the real lesson he wanted to draw is that, even so, realism is always ultimately preferable to instrumentalism because realists are *correct* in suggesting and defending unsupported conjectures. He accepted the conclusion of the instrumentalist's argument, but maintained that it is methodologically conservative, remaining at a superficial level:

> Any attempt to give a realistic account of the behaviour of the elementary particles is bound to be inconsistent with highly confirmed theories. Any such attempt therefore amounts to introducing unsupported conjectures in the face of fact and well-supported physical laws. This is the main objection which is used today against the theories of Bohm, Vigier, de Broglie, and others. It is similar to the objections which were raised, at the time of Galileo, against the idea that Copernicus should be understood realistically. ([1964a]: *PP*1, p. 195)

Feyerabend argued further that while Bohr gave the correct interpretation of quantum theory, the theory itself should be regarded as refuted, and should be replaced with one admitting of a realist interpretation. The right way for the realist to respond to the physical arguments for local instrumentalism is to find a way to defend unsupported theories against well-supported ones:

> He must then also revise the accepted *physics* in such a manner that the inconsistency is removed; i.e. he must actively contribute to the *development* of factual knowledge ... In addition he must offer methodological considerations as to why one should change successful theories in order to

be able to accommodate new and strange points of view. (*PP1*, p. 177)

A realistic alternative to the idea of complementarity is likely to be successful only if it implies that certain experimental results are not strictly valid. It therefore demands the construction of a *new theory*, as well as demonstration that this new theory is experimentally at least as valuable as the theory that is being used at present. (*PP1*, p. 193)

In arguing thus, Feyerabend manifested an excessively strong commitment to global realism, applying it at the expense of relinquishing even our best theories, if they have no satisfactory realist interpretation.[12] He did so because he wanted to show two things. First, he wanted to present realism as somewhat heroic or risky, in that it involves not only going beyond the evidence but also going *against* the evidence in certain cases, thus threatening to leave us with no unrefuted theory. Second, he wanted to show that realists will always be more keen on theoretical pluralism (or 'proliferation') than instrumentalists (or positivists generally). According to Feyerabend not only is there no harm in sticking to a realistic interpretation of a theory 'in the face of fact and well-supported physical laws' (despite the aforementioned risk of being left with no unrefuted theory), because this strategy has paid off handsomely in the past, but one is methodologically obliged to proceed in this manner, since the retention of 'refuted' theories, combined with the construction of new theories, is our guarantor of the theoretical pluralism which encourages scientific progress.

Feyerabend's argument for realism would be devastated if it could be shown either that non-realist positions are just as conducive to theoretical pluralism as realism is, or that theoretical pluralism, in the unrestricted form Feyerabend endorses, is an undesirable ideal for science. There is historical evidence that non-realists can be at least as pluralistic as realists, that they too can emphasize the proliferation of factually adequate but mutually inconsistent theories,[13] although for reasons different from the realist's. Giedymin argued that

[a]ncient and renaissance instrumentalism was associated with a pluralist view of scientific method. Its descriptive account of the method of science was based on a poly-theoretic model: one invents freely imaginative hypotheses to account for available observed data; one eliminates those which do not save the phenomena; however, this process is never conclusive and never terminates: there are always alternative hypotheses explaining facts equally well and predicting new facts with equal precision; the invention of new, mutually incompatible though observationally equivalent hypotheses is encouraged and their critical assessment is based on criteria transcending factual support; since incompatible, rival hypoth-

eses are, of course, not asserted, one is free to use several of them in research and see what results they yield. ([1976], p. 204)

In Giedymin's scheme, Osiander, Bellarmino, Duhem and Poincaré were 'modest' instrumentalists, rather than instrumentalists in Popper and Feyerabend's extreme sense. Giedymin's historical evidence for instrumentalist pluralism does not presuppose the Duhemian thesis that there was a tradition of extreme instrumentalism in ancient astronomy, and it is therefore immune from Musgrave's rival contention that Osiander and Bellarmino are better thought of as disappointed dogmatic realists.[14]

In fact, whether or not there was an illustrious tradition of instrumentalists in early astronomy, we might well believe on *a priori* grounds that instrumentalists can be more pluralistic than realists. This makes sense, because not all non-realists have the same degree of commitment as realists must have to the ideal of unity, the attempt to establish true universally-quantified statements.[15] Not only the kind of realism associated with an inductive conception of scientific method, but also more critical versions of realism must demand that, in the long run, theories from different branches of science should be unified and therefore mutually consistent. It stands to reason that those who do not demand that science provide a single, unified 'description of the world' are not forced to demand that scientific theories should be compatible with one another when taken as descriptions.

There is plenty of evidence for this in the history of non-realist views. Ernst Mach, perhaps the most extreme anti-realist, was explicitly pluralistic in his opposition to Max Planck's monist realism. Henri Poincaré, who was not an instrumentalist but a conventionalist, and Pierre Duhem (who was arguably neither) elaborated the aforementioned poly-theoretic model of scientific method. In his most instrumentalist moment, Poincaré noted that 'Two contradictory theories, provided that they are kept from overlapping, and that we do not look to find in them the explanation of things, may, in fact, be very useful instruments of research' ([1905], p. 216). His 'radical conventionalist' successors Edouard LeRoy and Kasimierz Ajdukiewicz were notorious pluralists who accepted an extreme thesis of incommensurability according to which different scientific theories simply could not be compared as to their content. Feyerabend's disregard of this desirable methodological feature of non-realism is especially inexcusable if we consider his repeated emphasis that realists *must* strive for a unified theory, and his recognition that non-realists are not subject to the same demand ([1958b], pp. 79–80). His desire to extrapolate scientific theories to all domains manifests a purely philosophical 'demand' with little support in science. (Of course it might be possible to enshrine this as a methodological principle of science, but (as the later Feyerabend would surely point out) actual science seems to involve no such principle). The early Feyerabend must be admonished not just for thinking of the realism/instrumentalism dispute as resolvable by a decision, but at the same time for trying to pass off 'scientific realism' as a component of science itself, rather than as a philosophical view about the nature of scientific theories.

In conclusion, we must rectify Feyerabend's conception of the realist/non-realist dispute in several ways: we should recognize more explicitly than Popper or Feyerabend that instrumentalism is not the only form of positivism, and that it may not be the most plausible form. We should also recognize that being a positivist is not the only way of not being a realist, and that a variety of positions lies between extreme instrumentalism and extreme realism. Just as sophisticated realists do not have to hold that all the non-logical terms in a scientific theory are referring terms, sophisticated non-realists can allow that scientific theories *can* be characterized as yielding factual claims, and therefore as true or false. Feyerabend's claim to have tackled the strongest form of instrumentalism is misleading: instrumentalism, like realism, becomes stronger the more moderate it is. The philosophical arguments for instrumentalism, which are arguments for global instrumentalism, may be indecisive or even, as Popper and Feyerabend thought, invalid (this issue has not been examined here). But the 'physical' arguments for instrumentalism are arguments only for local instrumentalism, and are definitely invalid, since they presuppose a generally realist position. Thus they are of no help to the global instrumentalist. Whether Feyerabend's grand methodological argument for realism succeeds, and thus defeats any non-realist alternative, including global instrumentalism, is yet to be finally determined. The historical evidence points against it, and non-realists seem to have a clearer rationale for theoretical pluralism than realists. We will later cast doubt on the unrestricted principle of proliferation that Feyerabend apparently needs to sustain his argument for realism, as well as on his argument for theoretical pluralism.

5

Theoretical Monism

5.1 The Myth Predicament

Traditionally, empiricism is the view that any substantive knowledge we have comes from experience and not from pure reason, revelation, intuition, authority or any other source. The classical British empiricist philosophers, for example, formulated their creed in this way. Popper tried to effect a decisive transformation in our conception of empiricism by arguing against the whole idea of 'sources' of knowledge. Feyerabend embraced this transformation:

> The theory of (scientific) knowledge which was developed in the 'thirties by Professor Karl Popper is the first example of what one might call a non-authoritarian epistemology . . . The reform of the theory of knowledge that accompanied the rise of modern science . . . did not eliminate the appeal to authority, which returned in the more abstract and depersonalised form of various 'sources of knowledge', such as Experience or Reason. Popper's contribution consists in eliminating the last element of authoritarianism, the idea that knowledge must have a foundation, and the correlated idea that its evaluation consists in investigating the manner in which it is related to this foundation. ([1965d], p. 88)

Part of Feyerabend's stated purpose was to develop and defend, as part of a normative epistemology, a 'disinfected', 'reasonable' or 'tolerant' empiricism according to which we ought not to admit theories which are wholly unconnected with experience. For Feyerabend, this involved accepting the 'principle of testability', which enjoins us to consider only theories which are testable, and exhorts us to maximize the testability of the theories we use ([1962a]: PP1, p. 45, [1964a]: PP1, p. 200). If we also accept falsificationism, which Feyerabend does at this point, we must restrict ourselves to considering only falsifiable theories. If we then endorse the equation of falsifiability with empirical content, as both Feyerabend and Popper do, we generate what Feyerabend calls 'the basic principle of empiricism', which allows us to consider only theories with high empirical content, theories which are maximally informative ([1962a]: PP1, p. 72). Empiricists argue that metaphysical theories fail to meet this demand. Feyerabend agreed. While he argued that metaphysical theories have a much more important role in science than positivists allow, he thought of them as proto-scientific, not scientific. But then, in an important and original line of thought, he tried to show that scientific theories can themselves, under certain circumstances, turn into untestable, and therefore metaphysical, theories. His objections to all forms of 'theoretical monism' (the restriction to using a single theory in any domain) are premised on the need to avoid such a situation. The main worry behind his objections was a feeling that these doctrines are conducive to the formation of what I shall call 'the myth predicament'. Feyerabend worried that such doctrines would not allow us to make the best use of our cognitive resources, especially our immense and unique capacity for wholesale reconceptualization, and further that they could easily lead, under unfavourable conditions, to the stagnation and stultification of science. Instead, he sought to persuade philosophers that being a good empiricist meant being tolerant about alternative theories, being a *theoretical pluralist*. He consistently opposed the more familiar empiricist demand to accept only theories closely connected with experience, as well as more familiar manifestations of that demand, such as sense-datum epistemologies, Newton's 'classical' empiricism, and the 'deductive-nomological' model of explanation.[1]

What is common to these doctrines, among others,[2] is that they open up the possibility of an unforced but irresistible enslavement of the mind. One of Feyerabend's most persistent themes is that truly comprehensive theories have an immense appeal, and an equally immense effect:

> [T]he influence, upon our thinking, of a comprehensive scientific theory or of some other general point of view, goes much deeper than is admitted by those who would regard it as a convenient scheme for the ordering of facts only. [S]cientific theories are ways of looking at the world; and their adoption affects our general beliefs and expectations, and thereby also our experiences and our conception of reality. ([1962a]: PP1, p. 45)

Feyerabend's line of thought starts with a thesis he credits to Immanuel Kant, later developed by Bohr. Kant had the idea that theories have a habit of turning into fully-fledged 'conceptual schemes'. Whereas Newton claimed to have derived his laws of motion from the phenomena, Kant claimed to be able to derive them *a priori*. Had this claim been taken seriously, Newtonian mechanics would have suffered this fate:

> A single theory in terms of which everything can be explained – as was the case with Newton's theory of celestial mechanics – is almost like a metaphysical system. It is impossible to free oneself from it; one can hardly imagine that it might be false; and one has no means of testing its correctness. ([1963e], pp. 96–7)

This is what Feyerabend is concerned to prevent: an aprioristic methodology (Kant's is, of course, a classic case) could lead to a myth predicament, a situation where one theory is regarded as forever unassailable. To the extent that empiricism (or any other philosophy) embodies such unacknowledged aprioristic elements Feyerabend finds it unacceptable. It will mask from us the fact that 'a doctrine that affects the mind of all can still *be* outrageously inadequate although it will of course not be easy to *show* its inadequacy' ([1964d], p. 251).

Bohr, too, assigned a special foundational role to classical Newtonian physics. For him, our belief in classical physics had influenced not just our thinking and our experimental procedures, but also our perception. Feyerabend endorsed this description of how a general physical theory can come to dominate our practices and perception.[3] But Bohr insisted further that our account of evidence in physics must be given in classical terms, that the classical 'forms of perception' must persist (see [1958a]: *PP*1, p. 22). Feyerabend pointed out that if this were so it would become increasingly difficult to imagine any alternative account of the facts. In one of his very first published writings, opposing Bohr's idea that classical physics might stand to quantum mechanics (and any future physics) as an *a priori*, whose laws comprise the only possible physics, Feyerabend called the background to this view a 'mistaken apriorism (which is, indeed, a concealed positivism)' ([1956b], p. 262). He later made an explicit connection between positivist theories of meaning, instrumentalism, and the myth predicament:

> The impression that every fact suggests one and only one interpretation and that therefore our views are 'determined' by the facts, this impression will arise only when . . . the relation of phenomenological adequacy is a one–one relation . . . [S]uch a situation arises whenever a fairly general

point of view was held long enough to influence our expectations, our language and thereby our perceptions, and when during that period no alternative picture was seriously considered. We may prolong such a situation either by explaining away adverse facts with the help of *ad hoc* hypotheses ... or by reducing more successful alternatives to 'instruments of prediction' which, being devoid of any descriptive meaning, cannot clash ... with any experience; or by devising a criterion of significance according to which such alternatives are meaningless. The important thing is that *such a procedure can always be carried out* ... This means that we can always arrange matters in such a way that either the principle of phenomenological meaning or the principle of pragmatic meaning will seem to be correct and that the stability-thesis correctly describes the relation of our knowledge to experience. But we can also choose the opposite procedure, i.e. we can take refutations seriously and regard alternative theories, in spite of their unusual character, as descriptive of really existing things, properties, relations, etc. In short, *although the truth of a theory may not depend on us, its form ... can always be arranged so as to satisfy certain demands.* ([1958a]: *PP*1, pp. 34–5)

This is Feyerabend's deeper and more important challenge, one respect in which even his early work goes beyond Popper: just as theories may be mere summaries of what is experienced, but only because we have settled for these undemanding products, so meaning may be determined by phenomenology, or by use, but only because we have temporarily arranged that it should be so. If we have made such an arrangement, we were wrong (by the lights of a reasonable methodology) to do so. But we can always revoke our decision and allow something else to determine meaning: the interpretation of our utterances, like the form of our knowledge, is malleable.

Such a line of thought is a double-edged sword. On the one hand it gives Feyerabend an extra set of resources. Any demonstration that his chosen semantic theory, or his favoured methodology, is not an accurate description of our linguistic or our epistemic practices will be irrelevant: he can retort that we have inadvisedly rigged the situation, and that adherence to a reasonable methodology would not have resulted in the (apparent or real) incorrectness of his semantic or epistemological theories. On the other hand, this same reasoning concedes that his own theories are not necessary truths. But when Feyerabend deploys the contextual theory of meaning he usually does so in a way which does not allow that the meaning of a term could be determined by anything *but* its theoretical context.

Like Popper, Feyerabend was interested in what distinguishes science from myth, and in what distinguishes 'primitive' societies from civilized ones. His most detailed discussion of myth occurs in his most Popperian paper, 'Knowledge without Foundations'. Scientific theories, according to Feyerabend, share certain features with myths. They are general and explanatory, they are counterintuitive and even counterinductive, for they go beyond and even contradict the evidence of the senses. Thus they are not like empirical generalizations, which merely repeat or summarize what is observed. Sometimes, in fact, scientific theories seem very much like the most undisciplined speculation. But going against the evidence of the senses provides means for investigating not only other theories but also observation reports, which are, after all, not absolutely trustworthy. Such theories are therefore far better instruments for the criticism and improvement of knowledge than empirical generalizations, which remain on the observational level.

What is it, then, which marks scientific theories off from myths? They cannot be distinguished by reference to experience as a source of knowledge, for myths have firm roots in experience, and can be supported by empirical arguments:

> Far from being a figment of the imagination that is clearly opposed to what is known to be the real world a myth is . . . a system of thought supported by numerous and very direct and forceful experiences, by experiences, moreover, which seem to be far more compelling than the highly sophisticated experimental results upon which modern science bases its picture of the world. . . . [T]here must be something amiss with the fairly popular idea that the distinction between a myth and a scientific theory lies in the factual basis of the latter. ([1961a], p. 23)

If we take the standard empiricist approach, we seem to be forced to admit that the only difference between myth and science is that science is the myth of today, whereas myths were the science of the past. Instead of pursuing such an approach, Feyerabend argues that we must look at the *psychological attitude* of the adherents as well as the *logical structure* of their theories.

Myths are self-supporting, Feyerabend says, because their logical structure offers us absolute truth.[4] They incorporate strategies for dealing with criticisms and with events that seem to refute them. If a theory, A, is faced with such an apparently refuting observation, O, an *ad hoc* proposal, B, which gives an account of O according to which O supports, rather than refutes, A may be put forward. But the certainty which results from such a procedure is entirely man-made: it is due to the fact that the theory's components have been related to each other in such a way that it is confirmed under all possible circumstances. The psychological attitude of those who believe in a myth is one of complete and unhesitating acceptance. Myths are treated by their adherents as absolutely true, certain, and infallible. The method of teaching appropriate to myths is indoctrination, they are preserved by fear, prejudice and deceit, and theories which differ from the accepted myth are regarded with suspicion. But the power of myths is not exhausted by these psychological factors; they can, as we have noted, stand on their own feet.

Despite the splendour of theories which offer, and deliver, absolute truth, there have always been thinkers who demanded something less impressive, but more human. Feyerabend followed Popper in identifying the pre-Socratic philosophers as the originators of the tradition of critical discussion which sets science apart from non-science, and sets civilized 'open' societies apart from uncivilized 'closed' ones: '[M]odern physics started . . . not as an observational enterprise *but as an unsupported speculation that was inconsistent with highly confirmed laws*' ([1963a], p. 5n). In contrast to the 'divine' Plato, who sought to protect human institutions by pretending that they have a supernatural origin, the Ionian philosophers taught that institutions have a human origin, and may therefore be imperfect and in need of change. They put forward bold conjectures that sought to explain the nature of the universe, criticized each others' conjectures, and made progress through the tradition of critical discussion in a 'critical community'. If this kind of philosophy has failed, Feyerabend thought, it is only because it has been taken over by dogmatism, not because of the boldness and generality of its hypotheses.

It is deeply ironic that Popper and Feyerabend, both of whom idolize science and seem to disparage philosophy, should take as their ideal a kind of intellectual activity that existed before the separation of science and philosophy. Both are nostalgic for a kind of inquiry which is most philosophical (even metaphysical), closest to philosophy done in 'the grand manner'. Popper does raise the question 'Is what the pre-Socratics were doing science, or just philosophy?', but he simply says that the answer does not matter. Neither he nor Feyerabend seriously addresses the question of whether mature science could really follow the pattern of pre-Socratic intellectual activity. Both of them import a distinctively philosophical *modus operandi* (scientific realism plus theoretical pluralism within the context of a 'critical community') from philosophy into science.

Feyerabend believed that myth predicaments would result from the enforcement of theoretical monism, and the accompanying destruction of critical communities. The first front on which he fought theoretical monism is apparently manned by the logical empiricists

Ernest Nagel and Carl Hempel. Feyerabend claimed that Nagel's account of reduction and Hempel's model of explanation are implicitly monistic, historically inaccurate, and that their enforcement would be undesirable. He argued that they lay down over-restrictive conditions on the introduction of new hypotheses. Let us, concentrating on the case of Nagel,[5] see what justice there is in these allegations.

5.2   Nagel on Science and Reduction

Nagel's model of science relies upon a distinction closely related to the one we saw Feyerabend draw between experimental generalizations and theories. According to Nagel, although observation reports are frequently couched in theoretical language, an experimental law has a life of its own, not contingent on the continued life of any particular theory that may explain it; such a law 'must be intelligible (and must be capable of being established) without reference to the meanings associated with it because of its being explained by [a particular] theory' (Nagel [1961], p. 87). Thus, Nagel's model assigns a certain autonomy to each of the various levels of the language of science (observation reports, experimental laws, high-level theories) at any time; Feyerabend therefore nicknamed it the 'layer-cake model' of scientific knowledge.

Any model of the structure of science also has implications for the historical development of science. Given any synchronic model, it must be possible to show in what scientific change consists. From Nagel's layer-cake model Feyerabend draws the following picture of scientific progress:

> The development of knowledge [on this model] consists in the accumulation of facts, and of theoretical layers. Science advances by internal improvement of each level, by addition of new facts on the observational level, by addition of new explanatory systems on top. There may of course be a considerable amount of change preceding the establishment of each single layer. Theories may have been formulated tentatively, and may again have been abandoned. But as soon as a theory is 'established in one area of inquiry' (Nagel), as soon as it starts pervading this area with its methods and its specific terminology, in the very same moment it assumes the independence of procedures and rules of usage that we have reported. The lower levels then actually become somewhat assimilated to the observational language and thus abstract conceptions assume a similar 'life of their own'. ([1966a]: *PP2*, p. 58)

This diachronic picture, claims Feyerabend, is wholly inadequate to account for scientific change, and furthermore would, if taken seriously by scientists, impose an intolerable straitjacket upon the development of science. His claim is that even an empiricist must recognize the unduly restrictive nature of the layer-cake model, which becomes manifest primarily in Nagel's ensuing account of theoretical reduction.

We should pause here to note some assumptions that Feyerabend is making. First, because he is most interested in reduction and explanation conceived of as diachronic relationships (relationships between successive theories in the same domain), he treats Nagel's model as a model of the development of science over time, not just as a model of its structure at a particular time. But reduction and, in particular, explanation, are often conceived of as synchronic relationships, relationships between statements within a single science, or between contemporaneous theories. By taking reduction and explanation as relationships between successive theories, Feyerabend creates the impression that logical empiricists were consciously concerned to lay down conditions on the development of science. And by taking reduction and explanation as conditions on scientific development, he creates the impression that these philosophers insisted that *any* successive theories in the same domain must be so related. This may be a little uncharitable, but we will let it pass. Even if they did not suppose that the relationships of theoretical reduction and deductive-nomological explanation are the only, omnipresent relationships between successive scientific theories, logical empiricists certainly supposed them to be widespread. Feyerabend's motivation is clear: he wants to show that *any* conditions on the development of science would be restrictive.

Reduction, according to Nagel, is an explanatory relation between theories (or 'sciences'), and a prominent structural feature of scientific knowledge: 'the phenomenon of a relatively autonomous theory becoming absorbed by, or reduced to, some other more inclusive theory is an undeniable and recurrent feature of the history of modern science' (Nagel [1961], p. 337). The existence of reductions is beneficial because, Nagel believes, insofar as an existing theory is reduced to a new theory it is saved from being an unqualified failure: we are vindicated in our past allegiance to it, we knew what we were talking about all along. Reductions fall into two classes: homogeneous ones are those where the reducing and the reduced theories share a conceptual apparatus, inhomogeneous ones are those where the reduced theory contains concepts not found in the reducing theory. Nagel thinks that although homogeneous reductions are common in the history of science, almost all philosophical interest attaches to the inhomogeneous ones. He then lays down formal conditions on reductions, two of which are as follows:

The Condition of Common Meanings   The reducing and reduced theories must have in common a large number of expressions that are associated with the same meanings in both theories. The expressions belonging to a theory possess meanings that are fixed by *its own* procedures of explication (see [1961], pp. 351–2).

The Derivability Condition   The experimental laws of the reduced theory must be shown to be *logical consequences* of the theoretical assumptions of the reducing theory. In cases where this is not possible because the laws of the theory to be reduced contain terms that do not occur in the theoretical assumptions of the reducing theory (that is, in inhomogeneous reductions), we must introduce linkages ('bridge laws') between what is signified by these terms and the properties or objects designated by the theoretical terms of the reducing theory. This ensures that the laws of the reduced theory are logically derivable from the laws of the reducing theory (see [1961], pp. 352–4).

This model of reduction has come under sustained attack over the last thirty years, an attack pressed home most forcefully by Feyerabend and his sympathizers. But Feyerabend does not directly attack these conditions themselves. Instead he argues that encapsulated both in these conditions on reduction and in Hempel's conditions on explanation are the following *implicit* conditions on the introduction of new theories:

The Condition of Meaning Invariance   Meanings must be invariant with respect to scientific progress; that is, all future theories should be phrased in such a manner that their use in explanations does not affect what is said by the theories or observation statements to be explained (see [1962a]: *PP*1, pp. 46–9),

and

The Consistency Condition   Only such theories are admissible in a given domain which either *contain* the theories already used in this domain, or which are at least *consistent* with them inside the domain (see [1962a]: *PP*1, p. 55).

Nagel's derivability condition is supposed to 'lead to' ([1962a]: *PP*1, p. 46) or imply ([1963a], p. 9; [1965a], p. 164) the consistency condition, while the condition of meaning invariance is allegedly implicit in his condition of common meanings. As long as we take it for granted that new and more general theories are always introduced to explain existing successful theories, Feyerabend thinks, this logical empiricist model of reduction insists that every new theory must satisfy the conditions of consistency and meaning invariance.

Feyerabend criticizes these conditions using what is to become a familiar, two-pronged strategy. Although he is officially committed to a strongly normative epistemology, and thus to the *irrelevance* of actual scientific practice, his critique of empiricism (indeed, of any methodology he disagrees with) proceeds, wisely, in two phases. He tries to show first that the methodology in question is not an accurate description of scientific practice; and then that its enforcement would be undesirable. In this way, his critique is stronger than the mere appeal to the methodological rules he prefers. Without the first phase, which he repeatedly stresses is of lesser importance, Feyerabend would probably never have moved away from normative epistemology to become known as one of the foremost 'historical' philosophers of science.

In the first phase of his critique of logical empiricism Feyerabend argues that the conditions of consistency and meaning invariance have not been enforced in the past history of science, that they are inadequate as descriptions of the relationship between successive theories. We shall deal here with the former condition, reserving discussion of the latter for the next chapter.

Feyerabend demonstrates that even the original derivability condition is not fulfilled in cases which Nagel regards as paradigm instances of theoretical reduction. For example, the 'reduction' of Galileo's law of free-fall to Newtonian physics is claimed by Nagel to be a case of simple homogeneous reduction, that is, 'no descriptive terms appear in the formulations of the Galilean science which do not occur essentially and with approximately the same meanings in the statements of Newtonian mechanics' (Nagel [1949], p. 291). But Feyerabend points out that the theories in question contain *incompatible* assumptions about the behaviour of material objects. Galilean physics assumes that the vertical accelerations of a body near the surface of the earth are constant over any finite interval. Newtonian mechanics, on the other hand, has vertical acceleration varying in inverse proportion to the distance between the object and the earth. As long as the value of h/r (h = height of object above ground level, r = radius of the earth) is positive, Galileo's law cannot be logically derived from the conjunction of Newtonian mechanics and a statement expressing the conditions valid inside the domain where Galileo's law applies. Thus the two theories, while the predictions they generate may be observationally indistinguishable, are not even, strictly speaking, consistent with one another: they transgress the consistency condition, as well as the derivability condition. No conventional or factual connections, no 'bridge laws', can patch up this situation by forging a non-logical connection between them. Feyerabend and his supporters conclude that

> [e]ven the favoured instances of apparently orderly accretions have turned out to be surprisingly disordered on close inspection. Indeed, there is not one example from the history of science that exactly fits the logical empiricist pattern of reduction, and some outstanding cases do not fit at all, force or bowdlerise how you will. (P. S. Churchland [1986], p. 281)

Feyerabend therefore offers us a choice:

> [w]e may either declare that the Galilean science can neither be reduced to, nor explained in, terms of Newton's physics; or we may admit that reduction and explanation are possible, but deny that deducibility, or even consistency . . . is a necessary condition of either. It is clear that the question as to which of these two procedures is to be adopted is of subordinate importance (after all, it is purely a matter of terminology that is to be settled here!) compared with the question of whether newly invented theories should be consistent with, or contain, those of their predecessors with which they overlap in empirical content. ([1962a]: PP1, p. 58)

Nagel responded by developing an account of reduction in which the reduced theory is approximately derivable from the reducing, in that the initial hypotheses may be reasonable approximations to the consequences entailed by the reducing theory. Feyerabend anticipated this move ([1962a]: PP1, pp. 58–9; [1965c]: PP1, pp. 111–13), claiming both that to resort to the idea of 'explanation by approximation' is to give up the derivability condition, and that the idea of approximation cannot be incorporated into a formal theory, since it contains essentially 'subjective' elements. For Feyerabend, this spelt bankruptcy for the descriptive ambitions of the orthodox theory of reduction.[6]

The second phase of Feyerabend's critique consists in asking not merely 'Is this how science proceeded?' but 'Would this be a *good* way for science to proceed?' He thinks that if Nagel's conditions apply retrospectively in the assessment of theories, we can turn them around to treat them as prospective (forward-looking) methodological rules. Cliff Hooker, in an excellent discussion of this issue, correctly suggests that superficial criticisms of Feyerabend are made here because most philosophers of science, deploying the distinction between the 'context of discovery' and the 'context of justification', assume that the rules for a prospective methodology of science need not be related to those for a retrospective assessment of science. But, he asks,

> [i]f we do indeed possess rationally defensible principles or rules for a retrospective assessment of the status of our scientific theories, ought they not also to be used as a guide for prospective methodology (and *vice versa*)? ... Feyerabend's (tacit) answer to this question is *Yes*. (Hooker [1972], p. 500)

And he argues in favour of Feyerabend's answer as follows:

> [S]uppose you have a set of rules which tell you, retrospectively, which of a bunch of theories is the best in the light of the evidence. Are you not then obliged, if you want to proceed *rationally*, to seek in the future just those theories which are supposed to be the best according to these criteria? ... Conversely, if you have a prospective methodology which tells you which kinds of theories to pursue it could hardly be credited as a *rational* methodology unless those theories were also accounted as the best. But in this case the criterion which tells you when you are pursuing the best kind of theory can be used retrospectively to tell you which of the theories you have pursued is the best. (pp. 500–1)

Feyerabend, turning Nagel's retrospective conditions for reduction around to use them as forward-looking methodological rules, argues that if the conditions of consistency and meaning invariance *were* enforced, this would put an end to scientific progress. He finds Nagel's conditions unacceptable when interpreted as methodological rules because he considers the resulting methodology inherently *monistic*.

Feyerabend's normative case against the derivability condition only works if we grant him the important assumption that more general theories are introduced in order to explain existing theories. Unless this assumption holds, accepting the derivability condition does not commit one to the consistency condition. We already deemed it uncharitable of Feyerabend to represent Nagel's conditions for reduction as if they were conditions on the introduction of any new theory. (Perhaps Nagel is at fault in failing to suggest other relationships between successive theories.) But if we do see science in this way, the resulting picture is one of old theories fitting within new ones, like a set of Russian dolls. That Nagel's model of reduction does imply the consistency condition becomes clear if we bear in mind that theoretical reduction is an explanatory relation between theories. Nagelian reduction demands that new theories explain old ones. But, everyone agrees, the consistency condition is unreasonable and, it has also been argued persuasively, has been endorsed by few practising scientists.[7] Nagel's model of reduction is thereby impugned. What is reasonably required of new theories is that they explain (or are at least consistent with) the relevant phenomena, not that they explain, or even agree with, old theories.

Feyerabend's critique of Nagel also has a broader point. Nagel's view of scientific progress is a case of cumulativism, according to which each new theory must incorporate major and essential features of its predecessor. As George Couvalis has argued, Nagel is a cumulativist in virtue of holding that many cases of reduction are normal steps in the development of a theory, and that even in the case of inhomogeneous reductions 'the reduction of one science to a second . . . does not wipe out or transform into something insubstantial or 'merely apparent' the distinctions and types of behaviour which the secondary science recognises' (Nagel [1961], p. 366). Cumulativism itself, let alone theoretical monism, may lead to the retention and petrifaction of error, or of inadequate conceptual resources. Whether it prevents us from ever revising the fundamental tenets of our conceptual scheme depends on what it is alleged must be retained.[8] Cumulativists started off very ambitiously, claiming that what must be retained is the whole existing theory. But their claim underwent a series of dilutions and modifications, successively becoming the weaker claims that what must be retained is: the original theory's truth-content, its confirmed parts, its theoretical laws and mechanisms, the reference of its central terms, etc. Larry Laudan, in an article that has achieved something of the status of a modern classic (Laudan [1981]), has argued persuasively that *none* of these demands are fulfilled. In the wake of Laudan's critique, cumulativists have become more modest still. The latest version of the doctrine is a return to Poincaré's 'structural realism', the view that the only thing that is (or must be?) retained from previous theories is some mathematical structure.[9] Feyerabend, of course, denies even this (see *PP1*, pp. 44–5, quoted below). And the thesis of structural realism, however weak it is, does seem open to empirical refutation. Cumulativism looks for all the world like a degenerating research programme. A modern Feyerabendian sums the situation up thus:

As historians and philosophers of science examined the dynamics of
science, they found it necessary to disensnare themselves from a background myth abetting logical empiricism – namely, the myth that science
is mainly a smoothly cumulative, orderly accretion of knowledge. (P. S.
Churchland [1986], p. 281)

5.3 Feyerabend's Anti-Reductionism

Feyerabend's positive views on reduction have been represented in
various ways. Kenneth Schaffner explained what he calls 'the PFK
paradigm', a conception supposedly derived from Popper, Feyerabend
and Kuhn, thus:

> The claim made in this paradigm is not that T_2 is derivable from T_1 in any
> formal sense of derivable, or even that T_2 can have its primitive terms
> expressed in the language of T_1, rather T_1 is able to explain why T_2 'worked',
> and also to 'correct' T_2. The relation between the theories is not one of strict
> deduction of T_2 from T_1. Nevertheless in certain cases one can obtain T_2
> from T_1 deductively: if one conjoins to T_1 certain contrary to fact premises
> which would in certain experimental contexts (relative to the state of a
> science) not be experimentally falsifiable, one can obtain T_2. (Schaffner
> [1967], p. 138)

This conception is recognizable from Popper, but Feyerabend should
not really be enlisted as an adherent. In 1962 he explicitly argued that
incommensurability precludes such reduction (although he held out
some hope that it might be rescued by the pragmatic theory of
observation). Even in later years, he certainly did not accept, as an
adherent of the PFK paradigm would, that a new theory could be
required to explain why an old theory (of the same domain) worked,
or even why it worked as well as it did:

> There is only *one* task we can legitimately demand of a theory and it is that
> it should give us a correct account of the *world*. ([1970a]: *PP*2, p. 159)

In fact, insofar as Feyerabend makes positive suggestions about the
relationship of reduction, he clearly conceives of it as outright *replacement*:

> What happens . . . when a transition is made from a theory T' to a wider
> theory T . . . is something much more radical than incorporation of the
> *unchanged* theory T' . . . into the context of T. What does happen is, rather,
> a *replacement* of the ontology (and perhaps even of the formalism) of T' by

the ontology (and the formalism) of T, and a corresponding change of the
meanings of the descriptive elements of the formalism of T' ... This
replacement affects not only the theoretical terms of T' but also at least
some of the observational terms which occurred in its test statements.
([1962a]: *PP1*, pp. 44–5)

He also stresses the *'subjective'* factors involved in theory-construction and theoretical reduction, arguing that which theory a scientist
suggests depends not only on the facts at his disposal, but also on the
tradition in which he participates, the mathematical instruments he
knows, his preferences, his aesthetic prejudices, and other subjective
elements. Whether or not this, sometimes called the 'replacement'
conception, should count as a conception of theoretical reduction (it
no longer involves theoretical explanation, but simply jettisons the
old explananda) is, as Feyerabend would say, merely a terminologi-
cal issue.

   While his critique of Nagel was part of a broad and popular move
away from restrictive models of theoretical reduction,[10] few drew
from the ensuing variety of models Feyerabend's conclusion that a
formal account of reduction and explanation is impossible. This
conclusion gave him, quite correctly, a reputation for being an 'anti-
reductionist'. It is not just that he lost interest in characterizing the
relationship of reduction or in finding any contentful relationship
between successive theories. For Feyerabend, the absence of a relationship of reduction between successive theories became a desidera-
tum, since it serves an important purpose. As we shall see, he came to
suppose that the formulation of incompatible theories facilitates the
task of critically evaluating our existing theory by allowing us to
identify, examine, test, and ultimately perhaps discard, its hidden
assumptions. Any methodology which rules out the introduction of
such alternatives is, for Feyerabend, insufficiently critical, over-
dogmatic, and 'monistic'.

5.4   Kuhn's Historical Case for Paradigm-Monism

When Feyerabend emigrated to the USA in 1959, having been offered
a post at the University of California at Berkeley, he met T. S. Kuhn
and, shortly thereafter, read Kuhn's book *The Structure of Scientific
Revolutions* in manuscript. In the next chapter we shall note what
Feyerabend called the 'pre-established harmony' between the two
which led them to produce, almost simultaneously, concepts of

incommensurability. However, there never was harmony between them over the issue of scientific methodology; indeed this subject constitutes their major forum of disagreement:

> [W]hile I thought I recognised Kuhn's *problems*; and while I tried to account for certain *aspects* of science to which he had drawn attention (the omnipresence of anomalies is one example); I was quite unable to agree with the *theory of science* which he himself proposed; and I was even less prepared to accept the general *ideology* which I thought formed the background of his thinking. This ideology, so it seemed to me, could only give comfort to the most narrow-minded and the most conceited kind of specialism. It would tend to inhibit the advancement of knowledge. And it is bound to increase the anti-humanitarian tendencies which are such a disquieting feature of much of post-Newtonian science. ([1970a]: *PP2*, p. 131)

Kuhn cannot be accused, as Nagel can, of cumulativism. He, more than anyone, insisted on the non-cumulative character of scientific 'revolutions'. But Feyerabend complained that while Kuhn gave good arguments for the existence of what he called 'paradigms', shared modes of scientific activity introduced by striking but open-ended achievements, he failed to motivate his own distinctive version of theoretical monism, the view that each domain of science at its best is dominated by a *single* paradigm. Feyerabend points out that Kuhn's case for this view, which I shall call 'paradigm-monism', is partly historical, partly functional.

The historical part consists in Kuhn's picture of the development of science, according to which science begins with a 'pre-paradigm' period in which researchers who can be called 'scientists' pursue research which yet does not constitute science. In such a period, there are no agreed canons of explanation, no fixed methodology, and no acknowledged scientific authorities, but rather a plurality of competing schools of doctrine, each deriving strength from its relation to a different metaphysical picture, but none having the upper hand. The activity of these researchers allows for unlimited disagreement and the criticism of each and every assumption. But this debate over fundamentals is directed towards other theorists, not towards nature. Even when there is a single theory of the domain, there is no single interpretation of that theory, no agreement on its achievements, methods, problems, or hopeful lines of solution. Observations which conflict with the theory are dealt with in an *ad hoc* way, or are simply ignored.

In science proper, long periods of success are punctuated with short periods of crisis. During the former, periods of what Kuhn provocatively calls 'normal science', a single paradigm dominates each field of mature scientific activity. Normal science is enshrined in the scientific textbooks of the time, for it is what must be taught to the new generation of scientific initiates. Because the paradigm represents a striking and open-ended achievement, induction into it counts as preparation for entry into the scientific community that holds it in regard. In that process of induction students learn the same models as their teachers. This ensures that there will be no fundamental disagreement between them. They will be committed to the same canons of explanation, the same 'ideals of natural order', the same rules and standards for scientific practice. Those commitments, Kuhn says, are the prerequisites for normal science. Thus the paradigm is not seriously questioned, but is taken for granted by those who work under it. Foundational issues are ignored, for this kind of scientific activity presupposes an at least partial 'closing of the mind'. But questions are now put to nature, rather than to other theorists. Problems are regarded as failures of application on the part of the scientists, 'puzzles' of no fundamental importance, rather than failures of the paradigm.

Periods of 'crisis', by contrast, are characterized by problems which are taken to reflect negatively on the paradigm itself. Before the transition to a new paradigm the old one is always perceived to be in trouble in some way. The trouble can take the form of the failure of normal science puzzles to turn out as the paradigm says they ought. This can multiply until we get a breakdown in puzzle-solving activity. One sign of crisis is the emergence of some features which characterize the pre-paradigm period, such as the proliferation of versions. Theories are proliferated in an attempt to deal with the problems plaguing the paradigm. In extreme cases there are as many versions of the theory as there are leading scientists working on it. Foundational issues are explored again, in the hope that they will yield new fundamental insight. The solid core of agreement which characterizes normal science is weakened. The crisis leads to a 'scientific revolution'.

Feyerabend summed up Kuhn's picture, which he later felt had become, in great part, the accepted picture of the development of science, in the slogan: 'Normal science is *monistic*; crises are *pluralistic*' ([1962c], p. 136). But he complained that Kuhn's picture was not just a description of science. He accused Kuhn of also insinuating an evaluation according to which normal science is desirable, and crises or revolutions are undesirable, or are desirable only as processes leading to improvement and normal science. 'Mature' science, he felt, had come to be identified only, and wrongly, with 'normal' science, and normal science was still the desired end-state: 'In short, it is believed that science is essentially normal science. Crises are embarrassments, periods of confusion which should be passed through as quickly as possible' ([1962c], p. 137). Feyerabend worried that the conservatism implicit in this sort of description of normal science had influenced both the historiography of science and the official methodology of scientists themselves. In other words, the philosopher's picture of methodology had corrupted methodology itself.

A case in point, according to Feyerabend, is the history of quantum mechanics, 'a marvellous example of the way in which philosophical speculation, empirical research, and mathematical ingenuity can jointly contribute to the development of physical theory' ([1962c], p. 138). At the turn of the twentieth century, when the grip generated by classical physics was forcibly relaxed, there were two very different reactions to the discovery that its foundations were infirm. Einstein, on the one hand, concluded that the official prevailing ideology of radical empiricism was wrong. He proceeded to invent a new theory, the special theory of relativity, which was susceptible of a realist interpretation. Those who developed the quantum theory, on the other hand, retained radical empiricism, concluding that it had not been correctly applied in classical physics, that classical physics itself contained metaphysical constituents. They therefore sought to construct a theory which would really satisfy the positivist ideal, a theory which did not go beyond experience, and which could not be given a realist interpretation. In doing so they were, according to Feyerabend, under the spell of an empiricism which, when applied to the history of science, has much the same monistic effect as the Kuhnian picture of scientific development. Their efforts resulted in the 'monolithic' quantum theory of the 1930s.

Feyerabend's methodological disagreements with Kuhn are many and various. He interprets Kuhn as holding that it is the existence of a puzzle-solving tradition that sets science apart from other activities, and he complains that this criterion of demarcation will not even exclude activities like organized crime and 'linguistic philosophy'! Sometimes, indeed, Kuhn denies the pre-paradigm stage scientific status because it does not have the features we associate with science today. While the people involved in this activity were recognizably scientists, their activity, he says, was 'something less than science' (Kuhn [1961], p. 355). Feyerabend calls this a 'purely semantic argument which *condemns* an activity because it is not *customary* to apply a certain word to it' ([1964d], p. 251). He therefore dismisses it as worthless,[11] identifying Kuhn's underlying failure on the problem of demarcation as a failure to consider the *aim* of science, and the question of whether 'normal science' will lead to that aim. Feyerabend's conception of rationality comes across clearly in his jibe that Kuhn surely cannot be saying that scientists have *no* aim, since this would make science irrational ([1970a]: *PP2*, pp. 133–4). But, as I suggested in chapter 1, talk about 'the aim of science' is problematic, and even if there is a single aim of science, it need not be the aim of scientists themselves.

Feyerabend's main disagreement with Kuhn, however, is focused primarily on 'normal' science, the period during which the dominant paradigm brings about a concentration of effort, the domain's foundational assumptions are taken for granted, and research is directed to the solution of more concrete and recondite problems. 'In receiving a paradigm', Kuhn says, 'the scientific community commits itself, consciously or not, to the view that the fundamental problems there resolved have, in fact, been solved once and for all' ([1961], p. 353). Normal science thus frees scientists from the discussion of fundamentals to engage themselves with tiny 'puzzles'. Feyerabend sees Kuhn as drawing the conclusion that

> [r]ecognition by society, institutionalisation, standardisation of teaching, in short, the arrival of many features which we today regard as characteristic of science . . . seem to be contingent on such a unity of doctrine and drastic restriction of debate. ([1964d], p. 244)

Feyerabend complains, correctly, that taken as an historical account, Kuhn's description of normal science is oversimplified, that normal science as Kuhn describes it is not even an historical fact. There *are* periods when one point of view reigns supreme, but they are not the only periods, and may not even be frequent. For example, towards the end of the last century there were *three* mutually incompatible points of view in physics: the mechanical point of view, the point of view connected with the development of the phenomenological theory of heat, and the point of view implicit in Maxwell's electrodynamics, developed by Hertz. The idea that this period was governed by a single all-embracing paradigm which absorbed physicists' attention to the exclusion of everything else is, Feyerabend submits, a caricature.

Kuhn admitted that mature science isn't an entirely monolithic enterprise, resembling rather a 'somewhat crazy, rambling structure' (Kuhn [1961], p. 387). But still he emphasized what he called the 'quasi-independent' character of the various parts of science, each of which is guided by its own paradigm and pursues its own problems.

Feyerabend countered by arguing, persuasively, that the crises which terminate the reign of a paradigm are often dependent on the active interaction of these inconsistent parts of mature science. It was the active interplay of the points of view mentioned above which led to the overthrow of 'classical physics', *not* the dedicated exploration of a single paradigm (which one?) or the independent exploration of the various paradigms. And he makes the point that Kuhn can give no explanation of where, if normal science is as monolithic as he thinks, competing theories could possibly come from. This part of Feyerabend's case seems ungainsayable.

But Feyerabend goes further, claiming that the troubles leading to Einstein's special theory of relativity could not possibly have arisen without an active exploitation of the tension between Maxwell's theory and Newton's mechanics, that it was not possible to use the phenomenon of Brownian motion for a *direct* refutation of the second law of phenomenological thermodynamics, but that the statistical-kinetic theory of matter had to be introduced from the start. These are important objections to paradigm-monism. But they ultimately rely on the impossibility of direct refutation and on the possibility of what Feyerabend calls 'indirect refutations'. These we shall deal with in chapter 7.

5.5   Kuhn's Functional Arguments

As Feyerabend and Nancy Cartwright have pointed out, Kuhn often provides *functional explanations* of scientific practices. *The Structure of Scientific Revolutions* has as a precursor a paper entitled 'The Function of Dogma in Scientific Research', in which Kuhn argues that

> Preconception and resistance seem the rule rather than the exception in mature scientific development. Furthermore, under normal circumstances they characterise the very best and most creative research as well as the more routine. Nor can there be much question where they come from. Rather than being characteristics of the aberrant individual, they are community characteristics with deep roots in the procedures through which scientists are trained for work in their profession. Strongly held convictions that are prior to research often seem to be a precondition for success in the sciences. . . Though preconception and resistance to innovation could very easily choke off scientific progress, their omnipresence is nonetheless symptomatic of characteristics upon which the continuing vitality of research depends. (Kuhn [1961], pp. 348–9)

Kuhn tries to show that the activity associated with mature science has far-reaching effects upon the development and substantiality of our ideas. This pedestrian activity, this concern for the solution of puzzles, the 'dogmatism of mature science', has, according to Kuhn, several (interrelated) functions.

First, it engenders commitment to a particular way of viewing the world and of practising science in it, a commitment which can be replaced but never merely given up, and which is actually constitutive of research by virtue of defining both the problems available for pursuit and the nature of acceptable solutions to them. But if commitment to a particular point of view is constitutive of research, so, it seems, are its corollaries, such as resistance to novelty, the authoritarian character of science education, and the unhistorical attitude which views the history of science as either filled with darkness or as already containing part of the present achievement.

Second, commitment to a paradigm is supposed to provide the scientist with a sensitive detector of the problems from which factual and theoretical innovations are teased. So although a 'quasi-dogmatic commitment' is a source of resistance and controversy, it is also instrumental in making the sciences the most consistently revolutionary of all human activities. Without a temporary closing of the mind, a quasi-dogmatic commitment to a certain point of view, mature science would be impossible. The occasional dogmatism exhibited by individual scientists is therefore a precondition for the success of science.

This second function of dogmatism is supposed to show that paradigm-monism is justified. It must be said that while Feyerabend clearly argues for both the existence and the desirability of theoretical pluralism, it is not so clear that Kuhn's arguments are supposed to be for paradigm-monism, and it is not clear whether their conclusion is that paradigm-monism is a historical fact, a desirable ideal, or both. Nevertheless, Kuhn does *seem* to defend not only the use of paradigms, but the monomanic concern with only one point of view, embodied in what he calls 'the exclusiveness of paradigms':

> At any time the practitioners of a given speciality may recognise numerous classics, some of them . . . quite incompatible one with the other. But that same group, if it has a paradigm at all, can have only one. Unlike the community of artists – which can draw simultaneous inspiration from the works of, say, Rembrandt *and* Cézanne and which therefore studies both – the community of astronomers had no alternative to choosing *between* the competing models of scientific activity supplied by Copernicus and Ptolemy. (Kuhn [1961], p. 352)

Kuhn's idea is that the more rigid the mental set shared by the scientists working under a paradigm, the more strongly will anomaly stand out, be recognized, and prompt work which either leads to its solution or, in failing to do so, initiates a 'crisis'. Feyerabend reports that Kuhn thought of this as a situation in which we get a close fit between theory and reality, a situation in which theory and reality 'confront one another directly'. Kuhn's second argument is supposed to be that only paradigm-monism will lead to such a situation, which is a precondition for progress in science. His argument is thus an appeal to the revolutionary nature of the developmental pattern of mature science:

> He defends such a procedure – and this leads to what I think to be the most important part of his philosophy – because he believes that its adoption will in the end lead to the overthrow of the very same paradigm to which the scientists had restricted themselves in the first place. If even the most concentrated efforts to fit nature into the current categories fail, if the very definite expectations created by these categories are disappointed again and again, then we are *forced* to look for something new . . . we are forced to do this by procedures which have established a close contact with nature, and therefore in the last resort by nature itself. This is the main reason why the rejection, by mature science, of the pre-paradigmatic battle of ideas is defended by Kuhn not only as a historical fact, but also as a reasonable move. ([1964d], p. 252; see [1970a]: *PP2*, p. 135)

Feyerabend agrees that these are both good arguments for paradigms, and that we need the 'dogmatism' of paradigms because they give scientists a guide. Nature is too complex to be explored at random, and the attempt to create knowledge cannot start from nothing. The researcher needs a paradigm in order to separate the relevant from the irrelevant and to indicate in what areas research will be most profitable.

Feyerabend also agrees with Kuhn that theories should not be eliminated at the sight of the smallest difficulty, that we must criticize both the theory and the statements describing this difficulty. They both endorse what Feyerabend calls the 'principle of tenacity', which urges us to stick to our theory despite the existence of *prima facie* refuting evidence and counter-arguments. It is rational to do so because theories are capable of development, can be improved, and may eventually be able to accommodate the difficulties which they were originally incapable of explaining.

But Feyerabend strongly contests the idea that Kuhn's arguments for paradigms point towards paradigm-monism. The psychological function of a single paradigm which Kuhn insists on, its function as a background against which the gestalt of an anomaly stands out, would be served just as well by a plurality of theories. He therefore complains that this dogmatism, the persistent commitment to a paradigm, simply does not add up to the monism Kuhn defends:

> [T]his plea for persistence, and this criticism of the theoretical butterfly who spends his entire life sitting now on the one, now on the other theory without advancing the cause of any is very different indeed from the combined historico-functional condemnation of a theoretical pluralism which we find in Kuhn's paper, and whose success would most probably be the end of 'the developmental pattern of mature science ... from paradigm to paradigm'. ([1964d], p. 253, quoting Kuhn [1961], p. 358)

Is it true that paradigm-monism, the preoccupation with 'puzzles', brings about a direct confrontation of theory and nature, and consequently progress through revolution? Feyerabend argues that it is not. He credits Kuhn with the idea that

> the close fitting between the facts and the theory that is a necessary presupposition of the proper organisation of the observational material can be achieved only by people who devote themselves to the investigation of one single theory to the exclusion of all other alternatives. For this *psychological* reason he is prepared to defend the dogmatic rejection of novel ideas at a period when the theory which stands in the centre of discussion is being built up. ([1962b], p. 277; *PP1*, p. 325)

This psychological conjecture is false, Feyerabend suggests, since many great scientists did more than just devote themselves to the development of a single theory.[12] He cites Einstein, Faraday and Newton as examples. Without such scientists it is doubtful whether there could ever be such a thing as a scientific *revolution* at all. Of course, Kuhn is thinking of the average scientist, not about the scientific genius. But, says Feyerabend, the inability of the average scientist to consider more than one theory (or even paradigm) at a time may be the result of inadequate scientific education, rather than being innate. Some type of commitment is certainly constitutive of research, but such commitment does not preclude a scientist's working on more than one paradigm at a time, or, more importantly, the existence, at any given time, of several competing paradigms, perhaps embodied in several different communities of scientists.

The points at issue between Kuhn and Feyerabend, therefore, are (a) whether there can be more than one paradigm at a time within a single science, (b) whether it is beneficial for there to be more than one such paradigm, (c) whether scientists can work on more than one paradigm (within a single science) at once, and (d) whether it is beneficial that they do so. Feyerabend's example of nineteenth-century physics provides an affirmative answer to the first and third questions, and goes some way towards answering the second and fourth affirmatively.[13]

Feyerabend has another strong case to make against Kuhn. He points out that since Kuhn accepts that refutations are impossible without the help of alternative theories, there can be a good functional argument from the necessity for scientific change to the desirability of theoretical pluralism. In his first published mention of Kuhn, Feyerabend tells us that he

> had the opportunity to consult as yet unpublished papers by Professor T. S. Kuhn in which the *noncumulative* character of scientific progress is illustrated very forcefully by historical examples. Despite some important and perhaps unalterable differences, the area of agreement between Professor Kuhn and myself seems to be quite considerable. One most important point of agreement is the emphasis which both of us put upon the need, in the process of the refutation of a theory, for at least another theory. As far as I am aware, this point has been made previously by K. R. Popper in his lectures on scientific method which I attended in 1948 and 1952. ([1962a], p. 32. See also [1995a])

In fact, although Feyerabend once conceded that Kuhn had given us 'a clear account of the function of alternatives in the development of science' ([1965c]: *PP*1, p. 108 n. 14), he later changed his mind, and accused Kuhn of misreading his own work:

> Kuhn has not only *admitted* that multiplicity [*sic*] of theories changes the style of argumentation. He has also ascribed a definite *function* to such multiplicity. He has pointed out more than once . . . that refutations are impossible without the help of alternatives. Moreover, he has described in some detail the magnifying effect which alternatives have upon anomalies and has explained how revolutions are brought about by such a magnification. He has therefore said, in effect, that scientists create revolutions in accordance with our little methodological model and *not* by relentlessly pursuing one paradigm and suddenly giving up when the problems get too big. ([1970a]: *PP*2, p. 140; emphases added)

So Feyerabend is claiming both that Kuhn's theoretical monism is descriptively or historically false (that there is no such thing as 'normal science' as described by Kuhn) *and* that it is methodologically undesirable. His arguments so far show that at the least Kuhn's model cannot possibly apply to all scientists and all scientific communities. There must be some who are capable of producing and developing new theories, for no number of anomalies in the old theory on its own adds up to a new theory.

Feyerabend's final argument against Kuhn is a reminder of what (in chapter 1) I called the 'ethical basis' of his philosophy. He asks whether Kuhn's picture, or his own, are attractive pictures of science, pictures which make the pursuit of science a worthwhile human aim. Because the sciences, within our culture, are surrounded by an aura of excellence which resists inquiry into their effects they are, ironically, removed from the domain of critical discussion. Yet

> why should a product of human ingenuity be allowed to put an end to the very same questions to which it owes its existence? Why should the existence of this product prevent us from asking the most important question of all, the question to what extent the happiness of individual human beings, and to what extent their freedom, has been increased? (PP2, p. 143)

What values should we use to probe today's sciences? According to Feyerabend in this 'humanitarian' mode, 'the happiness and full development of individual human beings is and has always been the highest possible value', and 'mature' science as Kuhn describes it is incompatible with these humanitarian values.

6

Incommensurability

6.1   The Condition of Meaning Invariance

In the previous chapter we saw that Nagel laid down, as a condition
on theoretical reduction, the requirement that reducing and reduced
theories must have in common a large number of expressions having
the same meanings. Feyerabend discerned in this a pernicious 'condition of meaning invariance':

> [M]eanings will have to be invariant with respect to scientific progress; that
> is, all future theories will have to be framed in such a manner that their use
> in explanations does not affect what is said by the theories, or factual
> reports to be explained. ([1963a], p. 10; [1965a], p. 164)

Is this a fair construal of Nagel's original clause? Feyerabend is
right to interpret Nagel as saying that reduction will not affect
the meanings of the reduced theory's 'primitive descriptive terms'
(the non-logical terms in its theoretical language and its observation-language), since Nagel does say that 'expressions peculiar
to a science will possess meanings that are fixed by its *own* procedures, and are therefore intelligible in terms of its own rules of
usage, whether or not the science has been or will be reduced to
some other discipline' (Nagel [1949], p. 301). But to interpret this
as a general injunction to the effect that all terms in future theories
must have the meanings they have (if any) in our current theories
is again to take him as suggesting that the relationship of reduction
is the only relationship that can hold between successive theories
in a given domain. In fact, Nagel explicitly treats theoretical

reductions as only one kind of scientific explanation.

The condition of meaning invariance should set bells ringing for us nevertheless: it is supposed to be a version of the stability-thesis. The meaning-invariance condition, says Feyerabend, is 'consistent with' the early positivism of the Vienna Circle: 'Their main thesis, that all descriptive terms of a scientific theory can be explicitly defined on the basis of observation terms, guarantees the stability of the meanings of observational terms' ([1962a]: *PP*1, p. 49. See also [1965a], p. 242 n. 82).

Feyerabend proposes to attack the condition of meaning invariance using his by-now familiar, two-pronged strategy. He aims to show first that it has not actually been satisfied in the history of science, and second that its satisfaction would be undesirable, inconsistent with a tolerant empiricism.

The condition of meaning invariance, he concedes, *has* sometimes been accepted in the history of science. Adherents of the Copenhagen Interpretation of quantum mechanics, for example, did subscribe to it. But this is an exception, born of mistaken radical empiricist ideology: the quantum theory of the 1930s is 'the *first* theory after the downfall of the Aristotelian physics that has been quite explicitly constructed with an eye both on the consistency condition and the condition of (empirical) meaning invariance' ([1963a], p. 12; emphasis added).

In other cases, those examples of radical conceptual change that have come to be known as 'scientific revolutions', and which Feyerabend treats as the decisive advances in the progress of scientific knowledge, the meaning-invariance condition was violated. Here, certain concepts in the new theory involved principles which were (on a 'realistic interpretation') incompatible with the principles of the older theory. Feyerabend's original example of this is taken from the medieval impetus theory, which was an attempt to remedy a problem in Aristotle's theory of motion. Aristotle's theory, according to which motion can only arise and persist through the continuous action of a force, could not account for the fact that projectiles continue to move in the absence of such force. The impetus theory came to the rescue with the hypothesis that the original mover transfers to the projectile an inner moving force (impetus) which is responsible for its continued motion. Feyerabend argues that the concept of impetus is not, as Nagel demands, explicable in terms of the theoretical terms of Newton's celestial mechanics, and that 'this is exactly as it should be, considering the inconsistency between some very basic principles of the two theories' ([1962a], p. 57).[1] Together with his other examples (thermodynamics ([1962a]: *PP*1, pp. 78ff), and mass in Newtonian and relativistic physics (pp. 81ff)), this is meant to refute the condition of meaning invariance, taken as a description of how science actually proceeds. Insistence on stability of meaning, Feyerabend concludes, would have made impossible some very decisive advances, such as the transition from Aristotelian physics to that of Galileo and Newton. This suggestion of inconsistency coupled with meaning variance already contains, as we shall see, the central problem with Feyerabend's semantic views.

For arguments against the desirability of meaning invariance, Feyerabend originally just refers us back to his arguments against the desirability of the consistency condition, supposing that the meaning invariance condition logically implies a special case of the consistency condition (see [1962a]: *PP*1, p. 82). His more substantial case against the desirability of meaning invariance, however, consists in a critique of empiricist methodology, and a positive case for a pluralistic methodology (to be described later).

So far, the meaning-invariance condition has been shown not to hold of certain important theoretical terms. Some empiricists concede this. But Feyerabend pushes further, extending his case to the realm of the observation vocabulary. He argues that scientific revolutions have changed the meanings of observation terms, even of terms from 'ordinary language'. Any resistance to this extension, Feyerabend thinks, must come from the idea that the meaning of observation terms is fixed by observational procedures (looking, listening, etc.). But what this idea overlooks is that

the 'logic' of the observational terms is not exhausted by the procedures which are connected with their application 'on the basis of observation'. . . [I]t also depends on the more general ideas that determine the 'ontology' (in Quine's sense) of our discourse. These general ideas may change without any change of observational procedures being implied. ([1963a], p. 16)

We have already cast doubt on this line of reasoning in chapter 2, where we showed Feyerabend failing to demonstrate that a term could suffer a change in meaning without its use changing in any way. Observation-terms do have (relatively) stable meanings, not because their meaning is fixed by invariant phenomenological features, but because it is fixed by their use, which is relatively impermeable to theoretical considerations. The introduction of a new scientific theory, although it may alter the meanings of observation terms, is not guaranteed to do so, as the contextual theory of meaning would have it. Feyerabend protests that 'invariance of *usage* in the trivial and uninteresting contexts of the private lives of not too intelligent and inquisitive people' does not indicate invariance of meaning ([1963a], p. 31). But his only support for this is his over-extended contextual theory of meaning.

6.2   The Thesis of Incommensurability

The most controversial consequence of the contextual theory is the incommensurability-thesis, over which a tremendous amount of ink has been spilt since 1962.[2] Feyerabend himself, notwithstanding later disclaimers,[3] had a lot to say about it. In the social sciences, its influence has been enormous. Within analytical philosophy, however, discussion has centred on whether it makes sense at all. In this chapter I shall give a history of the development of this thesis in Feyerabend's writings, up until the time of *Against Method*, drawing attention to some of its major problems.

Just as the contextual theory of meaning had its origins in Feyerabend's interpretation of Wittgenstein's later philosophy, so his thesis of incommensurability was inspired by a related source, the Cambridge philosopher Elizabeth Anscombe:

> On one occasion which I remember vividly Anscombe, by a series of skilful questions, made me see how our conception (and even our perceptions) of well-defined and apparently self-contained facts may depend on circumstances not apparent in them. There are entities such as physical objects which obey a 'conservation principle' in the sense that they retain their identity through a variety of manifestations and even when they are not present at all while other entities such as pains and after-images are 'annihilated' with their disappearance. The conservation principles may change from one developmental stage of the human organism to another and they may be different for different languages. I conjectured that such principles would play an important role in science, that they might change during revolutions and that the deductive relations between pre-revolutionary and post-revolutionary theories might be broken off as a result. (*SFS*, pp. 114–15. See also *KT*, p. 92)

Feyerabend recalls explaining this early version of incommensurability in Popper's 1952 seminar, and in a meeting in Oxford in the same year. The first occurrences of the idea of incommensurability (although not the term itself) in Feyerabend's published work are in early papers on complementarity,[4] where he remarked that

> a theory may be found whose conceptual apparatus, when applied to the domain of validity of classical physics, would be just as comprehensive

and useful as the classical apparatus, without coinciding with it. Such a
situation is by no means uncommon ... [For example], the concepts of
relativity theory are sufficiently rich to allow us to state all the facts which
were stated before with the help of Newtonian physics. Yet these two sets
of categories are completely different and bear no logical relation to each
other. ([1958b], p. 83; [1961b], p. 388)

And, just as there was a link between the stability-thesis and the
condition of meaning invariance, we are later told of a link between
'Thesis I' and incommensurability:

I interpreted observation languages by the theories that explain what we
observe. Such interpretations change as soon as the theories change. I
realised that interpretations of this kind might make it impossible to
establish deductive relations between rival theories. (*SFS*, p. 67)

This realization may well have begun as early as 1958, but its first
visible fruit appeared in 1962, when the term 'incommensurability'
itself was introduced.

When considering incommensurability we must recognize that,
although the thesis was developed in conversations between
Feyerabend and Kuhn in the early 1960s, Feyerabend's conception of
incommensurability is not the same as Kuhn's. Many critics, noticing
that the concepts are coeval, conclude that they are interchangeable.
But Feyerabend repeatedly insists that there is not a single shared
concept. Kuhn, he tells us,

observed that different paradigms (A) use *concepts* that cannot be brought
into the usual logical relations of inclusion, exclusion, overlap; and (B)
make us see things differently (research workers in different paradigms
have not only different concepts, but also different *perceptions*); and (C)
contain different *methods* for setting up research and evaluating its results.
According to Kuhn it is the *collaboration* of all these elements that makes a
paradigm immune to difficulties and incomparable with other paradigms.
Incommensurability in the sense of Kuhn ... is the incomparability of
paradigms that results from the collaboration of (A), (B) and (C). ([1977b],
pp. 363–4; *SFS*, pp. 66–7)

Feyerabend regarded (B) as having been refuted by empirical considerations. He insisted that his own incommensurability-thesis was
derived from problems in area (A) only: 'When using the term
"incommensurable" I always meant deductive disjointedness, *and
nothing else*' ([1977b], p. 365).

In 'Explanation, Reduction, and Empiricism' Feyerabend, starting
from the impetus theory example, constructed his first *general* charac-

terization of incommensurability, according to which the 'conceptual apparatus' of a new theory T is incommensurable with that of an older theory T' if and only if:

1 The primitive descriptive terms of T cannot be defined in terms of those of T';
2 There are no 'bridge laws' connecting the two sets of primitive descriptive terms which are correct and assertable consistently with T; and
3 The principles of the conceptual apparatus of T are inconsistent with those of T'.

These conditions are supposed to apply to many pairs of theories which have been used as instances of explanation and reduction: 'Many (if not all) such pairs turn out to consist of elements which are incommensurable and therefore incapable of mutual reduction and explanation' ([1962a]: PP1, p. 77). This is the widest claim for the scope of incommensurability Feyerabend ever made.

6.3 'On the "Meaning" of Scientific Terms'

This version of the thesis, and the contextual theory of meaning which licenses it, met with a mixed response.[5] Most germane to our interests is the heavy critical fire from Peter Achinstein, Dudley Shapere, Hilary Putnam and others. I shall concentrate here on Achinstein's acute but sympathetic critique, to which Feyerabend responded.

Achinstein's first fundamental objection was to the contextual theory of meaning. Theories of meaning are said to be *holistic* if they treat the meaning of a term as being derived from the meaning of some larger linguistic unit, such as a sentence, theory, or even a language. The contextual theory is holistic in virtue of treating the meaning of a term as a function of its surrounding theoretical context. But this entails that any change in a theory, however insignificant, would constitute a change in the meaning of all its descriptive terms. This view, which I shall call the 'extreme holistic thesis', has several unpalatable consequences. Every change in what we normally think of as a single theory would be tantamount to a substance-change, a change from one theory to a different theory. The history of 'a' theory would not consist in its being modified over time, but rather in there being a succession of theory-slices, none of which has any logical relations to one another. No two different theories could contain terms having the same meanings. Therefore, proponents of two different theories could neither contradict nor even agree with one another: every group of theorists would be talking only to themselves, of things about which no other theorists could say anything at all. This makes it difficult to see how two theories could be alternatives, rival accounts of the same domain. The contextual theory of meaning even makes nonsense of the concept of negation:

> According to this approach, if I assert p and you assert not-p, we are not and cannot be disagreeing, because the terms in my assertion are p-laden and so mean one thing, whereas those in not-p are not-p-laden and so mean another. Not-p, then, is not the negation of p. In short, negation is impossible! (Achinstein [1968], p. 93)

Feyerabend did not intend these consequences, but had given no clear account of what constitutes a change in the meaning of a theory's descriptive terms, and had failed to recognize that changes in belief (theory), and in meaning, come in many different shapes and sizes. If we tie the meaning of a term to the theories in which it occurs, we have to give criteria for when a change in belief constitutes a change in theory, and for when a change of theory constitutes a change in the meanings of the terms in question.

Achinstein also complained that the contextual theory ties the meaning of terms to a single feature (theoretical context), thus overlooking the facts that 'knowledge of many factors may be involved in understanding a scientific term and that, for some terms in a theory, quite a few of these factors may be known independently of the theory' (Achinstein [1964], p. 508). This is another area where a *use* theory of meaning might score decisively over holistic accounts like the contextual theory.

The second major problem Achinstein drew attention to was: what kind of inconsistency is adverted to in condition 3? A certain dilemma here has been the major obstacle to the presentation of any intelligible version of the incommensurability-thesis: either the inconsistency is logical inconsistency, or it is not.

Feyerabend had already chosen the first horn of this dilemma in asserting that the relation between certain pairs of theories (Newton's gravitational theory and Galileo's law of free fall, statistical thermodynamics and the second law of phenomenological thermodynamics, wave optics and geometrical optics), was logical: '[W]hat is being asserted here is *logical* inconsistency; it may well be that the differences of prediction are too small to be detectable by experiment' ([1963a], p. 13; [1965a], p. 168; and even AM^1 p. 36). But this does not fit with an incommensurability-thesis. If theories T and T' issue in logically inconsistent predictions, these must be expressible in some common (observation-)language, as a sentence of the form 'P & ~P', and this ensures that there is a logical or deductive relationship between the theories. These pairs of theories, at least, cannot involve meaning-changes, and cannot be incommensurable. If Feyerabend were to withdraw his claim that these theories are logically inconsistent, his case against the 'consistency condition' would be crippled. So he is under an obligation to give criteria of meaning-change which fit his other examples, but not these ones.

To embrace the other horn of the dilemma is to hold fast to the idea that the terms of a theory have their meanings fixed entirely by the principles of that theory, as the contextual theory of meaning suggests. In this case, it looks as if no two theories could ever be logically inconsistent with one another. To remedy this problem, the contextual theory of meaning should be amended in the way Achinstein proposes, by allowing for different kinds and degrees of dependence and independence that terms may exhibit with respect to theories in which they appear. This allows some different theories to be logically inconsistent with one another, while others are mutually inconsistent, but not logically so. Incommensurable theories will fall into the latter class. But if their inconsistency is not logical, what is to stop us using two or more incommensurable theories simultaneously? How can they be incompatible at all? The more radically two theories differ, the less plausible it becomes to think of them as rival accounts of a common domain or subject matter. If difference in belief exceeds a certain threshold, we lose our grip on the idea that the two theories are rival accounts of the same thing.

In 'On the "Meaning" of Scientific Terms' ([1965b]) Feyerabend replied to Achinstein using his favourite example of meaning-change, the change in the concept of mass from classical Newtonian mechanics to the theory of relativity. He had already argued that because classical mass is a property, whereas relativistic mass is a relation, the terms 'mass' in the classical theory and 'mass' in relativity theory mean very different things. He now definitively repudiated the extreme holistic thesis that all changes in theory are changes in meaning, and tried to lay down criteria for changes in meaning.[6] He asks us to consider three theories: T = classical celestial mechanics, T' = the general theory of relativity, and T* = T, modified by a slight change in the gravitational constant. The extreme holistic thesis would have it that even a change from T to T*, the kind which takes place regularly in the normal development of any scientific theory, involves a change in the meaning of all the theory's descriptive terms. Feyerabend, by contrast, insists that while T and T* are certainly different theories (their predictions do not coincide), the transition from one to the other leaves meanings untouched because there is no reason to assert that the different quantitative values of their forces is due to the action of *different kinds of entities*. This insistence reflects the intuitive difference in concepts which he earlier discerned: Newtonian and relativistic mass are 'different kinds of entities'. From these considerations he draws his criteria for stability of, and change in, meaning:

> This example shows that a diagnosis of stability of meaning involves two elements. First, reference is made to rules according to which objects or events are collected into classes. We may say that such rules determine concepts or kinds of objects. Secondly, it is found that the changes brought about by a new point of view occur *within* the extension of these classes and, therefore, leave the concepts unchanged. Conversely, we shall diagnose a change of meaning either if a new theory entails that all concepts of the preceding theory have zero extension or if it introduces rules which cannot be interpreted as attributing specific properties to objects within already existing classes, but which change the system of classes itself. ([1965b]: *PP*1, p. 98)

Most importantly though, Feyerabend asserts that such a diagnosis can be made unambiguously only if we have already chosen the kind of interpretation we prefer. Questions about the occurrence of a change of meaning, or of incommensurability, are relative to an interpretation, and thus have an unambiguous answer only *after* this decision has been made. Unsurprisingly, where matters of interpretation are concerned Feyerabend adopts precisely that form of realism which he proposed in 1958, encapsulated in 'Thesis I':

> This means regarding *theoretical principles* as fundamental and giving second place to the 'local grammar', i.e. to those peculiarities of the usage of our terms which come forth in their application to concrete and, possibly, observable situations. It is intended [*sic*] to subject the local grammar to the theories we possess *rather than* to interpret the theories in the light of the knowledge – or alleged knowledge – that is expressed in our everyday actions ... [W]e want to analyse, to explain, to justify and perhaps occasionally to *correct* the 'common knowledge' ... by relating it to new theoretical ideas, *rather than* to interpret such ideas as new ways of talking about what is already well-known. ([1965b]: *PP*1, p. 99; emphasis added)

Feyerabend justifies this 'epistemological realism' by saying not only that it lies behind the fundamental conceptual revolutions in the history of science, but also that it is demanded by a reasonable methodology. It is worth bearing in mind that this view, his version of scientific realism, is essential to establishing the existence of incommensurability. Feyerabend never relinquished the belief that the relationship of incommensurability between theories only exists if one has already taken the decision to interpret them 'realistically'. (Later, we shall catch him using the incommensurability-thesis long after his argument for realism has collapsed.)

Feyerabend's critics felt that his new criteria for change and stability of meaning raised at least as many problems as they solved. Achinstein reiterated that for a new theory to entail that its predecessor's concepts have no extension, the two must have common meanings. Shapere complained that for the criteria of meaning-change to work each theory must incorporate a unique and determinate set of classification-rules. If the rules were not determinate we would be unable to discern whether the system of classes had changed or whether there had merely been a change in the extension of the old classes. If the rules were not unique we might have a situation in which a change of meaning had taken place relative to one set of classification rules, while relative to another set no such change had occurred. But, he suggested, this would seem to be generally the case:

> One can, in scientific as in ordinary usage, collect entities into classes in a great variety of ways, and on the basis of a great variety of considerations ('rules'); and which way of classifying we use depends largely on our purposes and not simply on intrinsic properties of the entities involved. (Shapere [1966], in Hacking [1981], p. 51)

Feyerabend might well have thought that this, far from being an objection to his views, was one of his own points. Even though he repeatedly insisted that it is observation-sentences that need interpretation, and not theories, he also knew, in fact he stressed, that theories can be interpreted in different ways. And, as we have just seen, this is already taken into account by the thesis of incommensurability. But despite this latitude in interpretation, Feyerabend considered that there is an optimal way of interpreting theories, which he calls a 'realistic' interpretation. Scientific realists often appeal to the idea that the semantic ingredient in their view involves taking theories literally, or 'at their face value' (giving theories an interpretation which, as Wittgenstein might have said, is not an interpretation at all). Feyerabend seems to have thought that each theory, under a realistic interpretation, does indeed give a unique and determinate classification of entities. Under such an interpretation, a theory's rules of classification can simply be read off from it, and any theory with a different set of classification rules is *ipso facto* a different theory. This saddles Feyerabend with a very tight criterion of identity for theories (which conflicts with his usual practice). But, if it works, it means that a realistic construal of a theory determines that theory's ontological commitments, settling the question of which kinds of object that theory really countenances.

It is somewhat ironic that Feyerabend ultimately relies on the idea that each theory has what we might call a privileged interpretation. But in a curious way it reflects his increasing concern with the history of science. His real concern was that philosophers should interpret scientific theories in the way their adherents do:

> Of course, theories may be interpreted in different ways, they may be incommensurable in some interpretations, not incommensurable in others. Still, there are pairs of theories which *in their customary interpretation* turn out to be incommensurable. ([1977b], p. 365, n. 1; *SFS*, p. 68 n.; emphasis added)

The 'customary interpretation' is the one which scientists who defend the theory *use*, and which is taught to students. As Feyerabend put it in later work, 'it is the work of the physicists and not the work of the reconstructionists we want to examine' (*AM*[1] p. 253). Any reformulation, for example, a re-axiomatization, of a theory would not count as a 'customary interpretation', and its relation to other theories (including the question of their commensurability) would have to be decided anew. But the customary interpretation is privileged because it was that interpretation which was put forward, attacked, defended, and perhaps refuted in the course of scientific debate. *Post hoc* reformulations of a theory may be of interest, but their existence does not and should not enter into our explanations of scientific change, since they were not used by scientists in the debate at the time.

Unfortunately, there is no guarantee that the 'customary interpretation' of a theory always coincides with the interpretation dictated by realism. As Feyerabend himself noticed, the customary interpretation of quantum mechanics in the 1930s was hardly a realist one. In fact, he had argued more than once that this theory could not be given a realistic interpretation. But if a theory is interpreted by its defenders in any non-realist manner, it cannot be deemed incommensurable with any other theory. To some this will indicate that instrumentalism solves the 'paradox' of incommensurability; to Feyerabend it suggested that instrumentalism blinds itself to the existence of a relationship which can be perceived from a different and methodologically

better-motivated perspective. This uncomfortable position is an indicator of the half-way house between normative and naturalistic
epistemology in which Feyerabend at this time found himself. During the mid-1960s, he was unsure whether to give methodological or
historical considerations the upper hand.

Although Feyerabend fairly successfully distanced himself from
the extreme holistic thesis, and although his reply to Achinstein
should have put paid to the impression that he thought all pairs of
successive theories incommensurable, he fared less well in his reac-
tion to the accusation, often pressed by analytical philosophers, that
the incommensurability-thesis simply makes no sense. He distinguished between two versions of this objection. The first, stricter
version, says that theories which are logically inconsistent with one
another must share common meanings. Feyerabend conceded this,
and tried thinking of incommensurable theories as logically consistent, but as incompatible with one another in some other way. This
approach obliges its adherent to explain this concept of incompatibility in a way which makes it clear why we cannot use two or more
incommensurable theories simultaneously.

Several attempts have been made. Feyerabend himself proposed
that incompatibility of theories amounted to a *'lack of isomorphism*
with respect to certain basic relations plus consequences following
therefrom' ([1965c]: *PP1*, p. 116; emphasis added). Achinstein showed
clearly that lack of isomorphism is neither necessary nor sufficient for
two theories to disagree. First, theories with entirely different subject
matters can fail to be isomorphic, without disagreeing at all.
Feyerabend would reply that incommensurable theories must be
about the same subject matter. But we saw above that this is precisely
the requirement that his discussion of incommensurability fails to
illuminate. Second, Achinstein shows that theories can disagree only
in that one contains the negation of a proposition in the other, while
being fully isomorphic. Feyerabend might reply that this kind of
disagreement does not amount to the rich form of incompatibility that
interests him: logical inconsistency is only one mode of incompatibility.

Contemporary philosophers influenced by this phase of
Feyerabend's work, like Paul and Patricia Churchland, have pursued
this line of thought. Inspired by Feyerabend's materialism, and
taking their cue from developments in the brain sciences, they try to
move away from what they characterize as 'sententialist' models of
theory-construction and development, to find other, non-sentential
ways of representing theories and the relationships between them. So

far, they cannot be said to have explained how a lack of isomorphism is necessary or sufficient for theories to be incompatible.[7]

Another approach is pursued by Bill Newton-Smith, who proposes to come to the aid of Feyerabend (and Kuhn) by invoking something he calls 'pragmatic tension'. Two theories, T_1 and T_2, are in pragmatic tension if, when construed realistically, they lead us to posit the existence of different theoretical entities, states or processes, each set of which accounts for the phenomena:

> While there would be no logical inconsistency in adopting both T_1 and T_2, pragmatic considerations militate against this. For having adopted, say, T_1, we shall have an account of the phenomenon in question and there is no point in adopting T_2 as well. Indeed an application of Occam's razor gives a reason for not also adopting T_2. For nothing is gained by positing the existence of more than is needed to explain the phenomenon ... The theories T_1 and T_2 are in pragmatic tension in the sense that there is no need for explanatory purposes to adopt both, and adopting both would have the consequence of pointlessly bloating our ontology. (Newton-Smith [1981], p. 159)

While Feyerabend might not object to this, it is not what he needs. Ontological parsimony is not his motive for insisting that incommensurable theories are incompatible. What is more, as we shall soon see, he envisages a rough spectrum of kinds of incompatibility, stretching from the merely contradictory to the grossly incommensurable: we could not make any sense of this spectrum at all on the supposition that incompatibility is 'pragmatic tension', since pragmatic tension is a relation weaker than mutual contradiction.

On the second, looser interpretation, Achinstein's objection says only that competing theories must have common meanings. Feyerabend contended that this, despite its *prima facie* plausibility, is 'neither necessary nor desirable' ([1965b]: *PP1*, p. 103). He proposed to refute it by simply giving examples of pairs of theories which are both incompatible and incommensurable. The examples he used are classical celestial mechanics and the theory of relativity (see above), and classical celestial mechanics and the elementary quantum theory. It must be admitted that whatever strength the incommensurability-thesis possesses comes largely from its application to specific examples like these. Everyone accepts that the theories are competitors, and many are tempted by the thought that some concepts in the one theory cannot be formed within the other. The problem is that this thought precludes the most obvious *explanation* of their being competitors: their being logically inconsistent with one another.

In the later paper 'Reply to Criticism', Feyerabend offered a reformulation of the incommensurability thesis which appeals to the notion of meaning-change. Two theories, T and T', are here declared to be incommensurable if and only if: (a) The meanings of all their primitive descriptive terms are different, and (b) Any transition from T to T' (or *vice versa*) would involve a 'fundamental change' in rules ([1965c]: *PP1*, p. 115). Although merely a reformulation of the first version of the incommensurability-thesis, this generates problems of its own. Feyerabend explains the idea of the 'fundamental rules' of a theory thus:

> the rules (assumptions, postulates) constituting a language (a 'theory' in our terminology) form a hierarchy in the sense that some rules presuppose others without being presupposed by them. A rule R' will be regarded as being more fundamental than another rule R", if it [is] presupposed by more rules of the theory, R" included, each of them being at least as fundamental as the rules presupposing R". It is clear that a change of fundamental rules will entail a major change of the theory, or of the language in which they occur. (*PP1*, p. 114 n. 27)

As an explanation of the relation 'being a more fundamental rule than', this is multiply problematic. For a start it is circular, since the relation to be explained appears also in the explanation. Secondly, it presupposes that we can unambiguously count rules. But for this we need criteria of identity for rules, which are not forthcoming.

6.4 The Desirability of Incommensurability

Feyerabend's attempt to cast doubt on the idea that competing theories must have common meanings also has its normative dimension. He complains that such a principle is unreasonably restrictive, that it would prevent us from inventing and deploying the kind of alternative theories whose use would be methodologically optimal. In line with his 'principle of testability', a (wholly Popperian) methodological demand for the 'maximum testability of our knowledge' ([1965c]: *PP1*, pp. 104–5), Feyerabend suggests that the best methodology for science, or for any knowledge-gathering enterprise, is one which allows us to criticize our theories most severely:

> Criticism must use alternatives. Alternatives will be the more efficient the more radically they differ from the point of view to be investigated. It is bound to happen then, that the alternatives do not share a single statement with the theories they criticise. ([1963a], pp. 7–8)

The heart of his case is that a reasonable methodology demands the use of mutually inconsistent, partly overlapping, and empirically adequate theories. (This is Feyerabend's favoured 'pluralistic' methodology, to be examined in the next chapter.) Because of this, 'it thereby also demands the use of conceptual systems which are mutually *irreducible* ... and it demands that meanings of terms be left elastic and that no binding commitment be made to a certain set of concepts' ([1963a], p. 30). According to the principle of testability, major revolutions are preferable to small adjustments, since they lay open to criticism even the most fundamental assumptions. The idea that competing theories must have common meanings, by contrast, restricts the extent to which we are allowed to revise such assumptions.

There are several important ideas here, the first of which is the radicalism about conceptual change we drew attention to in chapter 3. The second is that meaning does not matter to science, that scientists simply take no notice of semantic matters. The following observations are all taken from a single year, 1965, the period during which Feyerabend was most concerned with meaning:

> Too great concern with meanings could lead only to dogmatism and sterility. Flexibility, and even sloppiness in semantical matters is a prerequisite of scientific progress! ([1965a], p. 181)

> [I]n the decision between competing theories 'meanings' play a negligible part. ([1965b]: *PP1*, p. 97. See also pp. 102, 103)

> [I]n philosophical arguments which are comprehensive ... appeal to meanings occupies only the most primitive initial stages and is quickly replaced by considerations of fact and method. ([1965c], pp. 258–9)

> As far as I am concerned, even the most detailed conversations about meanings belong in the gossip columns and have no place in the theory of knowledge. This is true even in those cases where meanings are invoked to force a decision about some different matter. For even here their only function is to conceal some dogmatic statement which would not be accepted, if presented by itself, and without the chatter of semantic discussion. ([1965c]: *PP1*, p. 113. See also p. 114 n. 27)

These remarks have sometimes been interpreted, by sympathizers such as Hacking and Rorty, as indications that Feyerabend put forward no theory of meaning. This reading does not sit well with the coexistence of equally heartfelt expressions of the contextual theory of meaning. That meaning does not matter to scientists would not imply that it (and its offspring, such as incommensurability) can be ignored by those giving a philosophical account of science.[8]

Feyerabend's third idea in this line of argument is that not all alternative theories are equally suited for the purposes of criticism, that there is a spectrum of degrees of critical power, at one extreme of which is a relationship between theories which is the perfect embodiment of the critical ideal. Some theories are merely *ad hoc* variations on others: they contradict one another, but they do not differ in their fundamental conceptual apparatus: they are semantically commensurable. Others, however, stand in a stronger, more adversarial relationship than merely contradicting one another. For two theories, T and T', to be what Feyerabend calls 'strong alternatives',

1 they must contradict each other over some observation statement;
2 T', the alternative to our existing theory T, must 'contain assertions over and above the assertion (the prediction) which leads to the contradiction';
3 these additional assertions 'must be connected with the contradicting assertion more intimately than by mere conjunction';
4 there ought to be some independent reason to accept T', e.g. evidence in its favour; and
5 T' 'should also be able to account for the earlier successes of the criticised theory [T].' ([1965c]: *PP*1, p. 109).

For a new theory to count as a strong alternative to our existing one it must be coherent, well-confirmed, not merely an *ad hoc* variant on our existing theory, and it must be able to account for the observational success of its predecessor.

Theory-testing using 'strong alternatives' sounds exciting, but problematic. Can one test the assumptions of a theory by introducing a new theory, none of whose terms mean the same as those of the existing theory? Since the new theory cannot even assert what the old one denies, or vice versa, such a 'test', a 'competition' between incommensurable theories, must be an exercise in which the protagonists systematically talk past one another, a dialogue of the deaf. But this is not an objection to Feyerabend, since he (like Kuhn) agrees that comparison of high-level scientific theories (or paradigms) really *is* like this. For Feyerabend, in fact, this is an objection to accounts of science which insist that theory-testing or theory-comparison must respect semantic relationships, not to the process of theory-comparison as *he* conceives it, since that process, as we have just seen, rides roughshod over semantic matters. As long as there are some (genuinely non-semantic) criteria for choosing between incommensurable theories, criteria that scientists could (and, preferably, do) use, Feyerabend can argue that the procedure he has in mind makes sense.

The *real* problem with Feyerabend's requirements for strong alternative theories is that he wants to say that the primitive descriptive terms of two strong alternative theories are *incommensurable* with each other. How can this possibly be? Strong alternatives *contradict* each other at the observational level, but incommensurable theories *cannot* be logically inconsistent, for '*[n]ot a single primitive descriptive term of T can be incorporated into T''* ([1965c]: *PP1*, p. 115). The impression created by these sections of 'Reply to Criticism' is that incommensurable theories have been shown to be better suited for the purposes of criticism than merely contradictory pairs. But no such thing has been achieved. Despite Feyerabend's explicit avowal that Achinstein's objection was 'entirely correct', what he actually attempts to show is that incommensurable theories are a special class of mutually contradictory theory-pairs, but this, as he knew, is impossible.

6.5 Comparing Incommensurable Theories

So far, believers in incommensurability have not made their case, for they have not yet spelt out the way in which incommensurable theories are incompatible with one another. But this does not mean that their task is hopeless. In this section I shall focus on a problem which persists even if a suitable characterization of incommensurability can be found.

This is the problem pertaining to theory-testing. As already indicated, Feyerabend's 'pluralistic test model' positively demands that incommensurable theories be compared with one another. This gives the lie to those commentators who claim that incommensurability entails (or even constitutes) incomparability.[9] Although Feyerabend did once say that 'incommensurable theories may not possess any comparable consequences' ([1962a]: *PP1*, p. 93), this remark should be interpreted as referring to semantic comparability only. His more considered verdict is that while succeeding theories can be evaluated only with difficulty and 'may be altogether incomparable, at least as far as the more familiar standards of comparison are concerned . . . *they may be readily comparable in other respects*' ([1970a]: *PP2*, p. 152; emphasis added). Indeed, he insists that he 'never inferred incomparability' from incommensurability (*SFS* p. 68; [1977b], p. 365). For Feyerabend, incommensurability precludes (a) any judgement of relative verisimilitude, (b) any comparison of content-classes, and (c) any judgement of approximate truth. Since several popular comparison measures are based on such criteria, those who defend them rightly feel threatened by incommensurability: they insist that if incommensurability were a fact, there would be no objective theory-comparison at all, and 'subjectivism' would have won the day. Feyerabend notes that this conclusion is a problem only for those whose account of science requires that if theories are to be assessed objectively, they must be semantically comparable. When Laudan, for example, complained that Feyerabend and Kuhn 'began to despair about the possibility of any objective yardstick for comparing different theories and suggested that theories were incommensurable and thus not open to objective comparison' (Laudan [1977], p. 143), Feyerabend replied that

> [t]his suggests that we wanted to compare theories, were misguided by some feature of science into believing that a comparative evaluation was impossible, and joylessly published the disagreeable consequence. A look at our work reveals an entirely different story. What we 'discovered' and tried to show was that scientific discourse *which contains detailed and highly sophisticated discussions concerning the comparative advantages of paradigms* obeys laws and standards that have little to do with the naïve models which philosophers of science have designed for that purpose. *There is* theory comparison, even 'objective' comparison, but it is a much more complex and delicate procedure than is assumed by rationalists. ([1981]: *PP2*, p. 238)

He continued:

> None of the writers who defend 'objective' standards has explained what the word means . . . Popperians occasionally connect objectivity with truth (in Tarski's sense) and call comparisons 'objective' only if they are based on a comparison of truth-content. Incommensurability rules out such a comparison. For a Popperian the remaining standards (*and there are many standards left*) are 'subjective', which is the reason why I call them 'subjective' in my criticisms of Popperians. Laudan takes [these passages] as indicating that *I myself* hold them to be 'subjective' . . . and he assumes that I apply incommensurability to *all* means of comparison, not only to means that depend on content. (*PP2*, p. 238 n. 17)

When Feyerabend conceded that logically inconsistent theories must share common meanings, he also realized that the comparison of incommensurable theories must be based on methods which do not depend on their semantic comparability. Such methods, he went on to claim, are readily available. We can compare incommensurable theories by:

1 using the 'criterion of predictive success' associated with the Pragmatic Theory of Observation ([1962a]: *PP1*, p. 93; [1965a], pp. 214–
   17);
2 devising a kind of 'crucial experiment' in which one of two rival
   strong alternatives is confirmed ([1965c]: *PP1*, p. 116);
3 comparing the efficiency with which each theory reproduces the
   'local grammar' of sentences which are directly connected with
   observational procedures (ibid.);
4 constructing a model of one theory within another and rejecting the
   former if the model violates highly confirmed laws (ibid.);
5 inventing 'a still more general theory describing a common background that defines test statements acceptable to *both* theories'
   ([1965a], pp. 216–17);
6 conducting an internal examination of the two theories in order to
   determine whether one has a closer connection to observation than
   the other (ibid.);
7 using various *formal* criteria such as linearity, coherence, predictive
   power, the number and character of approximations used (*SFS*, p.
   68, n. 119; [1977b], p. 365, n. 2);
8 using more *informal* criteria such as 'conformity with basic theory
   . . . or with metaphysical principles' (ibid.).

Some of these criteria, Feyerabend concedes, *are* 'arbitrary' or 'subjective', in the sense that 'it is very difficult to find wish-independent
arguments for their acceptability' (ibid.). In fact they are a mixed bag,
by no means uniformly important or plausible. While Feyerabend
devotes considerable attention to some of them, others are merely
mentioned, never to be examined in detail. Even a fairly cursory
examination will reveal that some of them should not have got off the
drawing-board.

1 The most important suggestion is the first. It is important because
predictive success is an empirical virtue which scientists certainly do
take into account in making theory-transitions. The 'criterion of
predictive success' makes use of the fact that

> our theories, apart from being pictures of the world, are also instruments
> of prediction. And they are good instruments if the information they
> provide, taken together with information about initial conditions charac-
> terising a certain observational domain D_o would enable a robot without
> sense organs, but with this information built into it, to react in this domain
> in exactly the same manner as sentient beings who, without knowledge of
> the theory, have been trained to find their way about D_o and who are able

to answer, 'on the basis of observation', many questions concerning their surroundings. This is the criterion of predictive success, and it is seen not at all to involve reference to the *meanings* of the reactions carried out either by the robots or by the sentient beings . . . All it involves is *agreement of behaviour*. ([1962a]: *PP1*, p. 93)

According to the Pragmatic Theory of Observation, 'p' is an observation-statement if and only if a well-conditioned and appropriately prompted observer who is situated in front of an appropriate object will respond to interrogation with 'p'. This, a behaviouristic substitute for the characterization embedded in the rival 'Semantic Theory of Observation', is an attempt to bypass the semantic aspects of observation-statements (which are, in the case of incommensurable theories, incomparable, and therefore cannot afford a basis for tests) in favour of their causal or pragmatic features. And, as I suggested in chapter 3, it is an attempt to see human observers as measuring-instruments.

A test of a theory is, according to this conception, a matter of comparing the reactions of the suitably programmed observer with the 'reactions' of the theory itself, the sentences which it predicts. The theory passes the test if and only if its deliverances successfully mimic those of the observer. The meanings of the predicted observation-sentences do not enter into the test procedure, since it is sentences and not statements which are compared. Thus this test procedure is adjudged to afford a means of comparing the virtues of even the most radically incommensurable theories.

This suggestion has come under plenty of fire.[10] Shapere conducted a sustained attack on the idea that the Pragmatic Theory of Observation could help us decide between rival theories of any kind. The gist of his critique is that, since uninterpreted observation-statements convey no information at all, they 'cannot convey information which would serve as a basis for "removing" a theory' (Shapere 1966, p. 48). Feyerabend suggests that 'an *acceptable* theory . . . has an inbuilt syntactical machinery that *imitates* (but does not *describe*) certain features of our experience' ([1965a], pp. 214–15). But why should such mimicking constitute a virtue of a theory? Without semantic content, observation-sentences seem to have no properties in virtue of which they are relevant, either positively or negatively, to the acceptability of theories.

2 The second criterion of theory-choice suggested by Feyerabend is also a substitute for the crucial experiment. The suggestion is that we 'confirm one of two incompatible theories. If the confirmed theory

and its rival are strong alternatives, then the rival will have to go'
([1965c]: *PP*1, p. 116). We have already seen that 'strong alternative'
theories are not incommensurable, since they contradict each other
over the truth-value of some observation-statement. But forgetting
the parenthetical requirement that the two be strong alternatives, and
concentrating on merely incommensurable theories, this suggestion
seems acceptable. Everything depends upon the process of confirmation. Theory-comparison can only work along these lines if we have
a measure of confirmation which is common to the theories. Such a
measure will be purely syntactic. Harold Brown, a commentator
sympathetic to Feyerabend, has noted that

> [i]t is one of the intriguing ironies of recent discussions in the philosophy
> of science that logical empiricists, who are among the most vociferous
> critics of the incommensurability thesis, have also sought purely syntactic
> confirmation theories as the ideal tool for theory comparison. Such tech-
> niques, were they available, would provide a straightforward means of
> comparing theories irrespective of their content, and would thus permit
> the comparison of incommensurable theories. (Brown [1983], p. 28, n. 19)

3 The third procedure is more complicated. Recall that Feyerabend
defines the 'local grammar' of a statement as 'that part of its rules of
usage which is connected with such direct operations as looking,
uttering a sentence in accordance with ostensively taught (*not* defined) rules, etc.' ([1965c]: *PP*1, p. 116 n. 32). He then suggests that two
theories can be compared if both can reproduce the local grammar of
sentences which are directly connected with observational procedures:

> In this case the utterance of one of the sentences in question in accordance
> with the rules of the local grammar, or the utterance of a local statement,
> as we shall say, can be connected with *two* 'theoretical' statements, one of
> T, and one T' respectively (theoretical statements, corresponding to a given
> local statement S will be called statements associated with S within the
> theory in question). We may now say that the empirical content of T' is
> greater than the empirical content of T, if for every associated statement of
> T there is an associated statement provided by T', but not *vice versa*. And we
> may also say that T' has been confirmed by the very same evidence that
> refutes T if there is a local statement S whose associated statement in T'
> confirms T' while its associated statement in T refutes T. (*PP*1, p. 116)

This measure of empirical content, the number of 'associated statements' a theory contains, is supposed to be an index of the closeness
or directness of the theory's connection to observation. Its main
problem is the relationship of 'connection' or 'correspondence' as-

serted to hold between 'local statements' and high-level theoretical statements. A single statement can only 'correspond' to two theoretical statements if it bears some relation of theoretical relevance to them (both), but if it does so there must be a semantic connection between the theoretical statements themselves, they must be commensurable. Unless the original 'local statements' are semantically relevant to the theoretical statements, it is difficult to see how they could bear on those statements in any interesting way. Moreover, how can an 'associated statement', which is a theoretical statement corresponding to a given local statement, confirm or refute a theory (of which it is a part)? The envisaged procedure looks murky.

4 Next it is proposed that in order to compare two incommensurable theories, T and T', we should

> construct a model of T within T' and consider its fate . . . Within physics the construction, in T' of a model T* of T usually is not possible without violation of highly confirmed laws. This, then, is sufficient evidence for the rejection of T* and, via the isomorphism, of T. ([1965c]: *PP*1, pp. 116–17)

Surprisingly, this criterion embodies a very conservative attitude towards new theories. The new theory T is removed not because it is inadequate in itself (it may be highly confirmed) but because it lacks a structural identity with or similarity to an existing theory. Feyerabend should be the first to insist that this is an unreasonable constraint on new theories. He seems here to have entirely lost sight of his own powerful critique of methodological conservatism. Furthermore, how can two theories which have models in common be incommensurable in the first place? Establishing the existence of such an isomorphism would, we are told, give us reason not to reject the new theory, but how could we then ever move from one theory to another which is incommensurable with it? The suggested procedure restricts theory-transitions to transitions between commensurables!

5 With the next suggestion we can have little patience. We are told that in cases where two theories do not share a single observation-statement we can nevertheless invent a 'still more general theory describing a common background that defines test statements acceptable to *both* theories' ([1965a], p. 217). Shapere gives this suggestion the execution it deserves

> [T]his third theory is still a different theory, and even though it contains a subset of statements which *look* exactly like statements in the two original

theories, the meanings of those statements in the new metatheory will still be different from the meanings of the corresponding statements in either of the two original theories . . . Thus, the same problems arise concerning the possibility of comparing the metatheory with either of the two original theories as arose with regard to the possibility of comparing the two original theories with one another. (Shapere [1966], p. 45)

This suggestion for comparing theories never raised its head again in Feyerabend's work.

6 What about a comparison based upon an internal examination of the two theories? 'The one theory might establish a more direct connection to observation, and the interpretation of observational results might also be more direct' ([1965a], p. 217). First, what does having a more direct connection to observation mean? Secondly, even if we could establish that T had a more direct connection than T', so what? Would *that* be any reason for preferring T? Certainly not for a liberalized empiricist like mid-period Feyerabend. We do not need to agree with Shapere that this procedure is unintelligible because Feyerabend holds that 'each theory defines its own facts or experience' (Shapere [1966], p. 45) to agree that it is useless.

7 Criteria such as linearity and coherence, not to mention predictive power (already dealt with), and the number and character of approximations used, are deployed by scientists in justifying theoretical changes. But in order to decide whether such criteria will apply to the case of incommensurable theories, Feyerabend must specify how they are to be gauged, in terms of measures which do not involve semantic considerations.

8 'Informal' criteria also have a place in the historical explanation of scientific change. They are among the criteria scientists appeal to, although those influenced by positivism claim to renounce them outright. Of course, drawing attention to them supports the impression that advocates of incommensurability can appeal only to 'irrational' or 'subjective' factors in theory-comparison.

This impression should be dispelled. Feyerabendian incommensurability, at least, does not entail or involve incomparability.[11] In fact, contrary to what critics like Putnam and Shapere have thought, the incommensurability thesis does not even entail *relativism*. Relativism, in philosophy of science, is the view that different scientific theories (or paradigms, or styles of theorizing) cannot be assessed by a single set of truth-relevant criteria. But, as we have just seen, some of the criteria with which scientists can compare incommensurable theories *are* relevant to assessing the truth of the theories in question. Only when taken together with an incomparability-thesis which would deny us the use of such criteria does incommensurability entail relativism (and whether relativism amounts to 'irrationalism' is a further question).[12]

This conclusion runs against a strong current of Feyerabend interpretation, according to which Feyerabend was a relativist, knowingly or not, during the phase of his work I am now discussing (roughly, the early and mid-1960s). This is a myth, originally generated by critical commentators, then reinforced when Feyerabend himself finally did succumb to relativism. It is a myth fostered also by Feyerabend's later suppression of passages from his own earlier work. Not only was Feyerabend *not* a relativist during this phase, but in the two or three places where he does discuss relativism he makes clear his *anti-*relativist orientation. For example, in attacking philosophers of science of the 'historical' school, for whom 'actual scientific practice is the material from which [to] start, and a methodology is considered reasonable only to the extent to which it mirrors such practice' ([1962a], p. 60), Feyerabend argues against the idea that different subjects, having a 'logic' of their own, cannot be judged according to a set of standards embodied in a single scientific methodology. Modern science, he insists,

> is the result of a conscious criticism of the theses propagated and the methods employed by the great majority of scholastic philosophers. For the thinker who demands that a subject be judged 'according to its own standards', such criticism is of course impossible; he will be strongly inclined to reject any interference and to 'leave everything as it is'. It is somewhat puzzling to find that such demands are nowadays advertised under the title of philosophy of science. Against such conformism it is of paramount importance to insist upon the normative character of scientific method. ([1962a], pp. 61–2)

His reasoning in these sections (not reprinted in his *Philosophical Papers*) is absolutely explicit: relativism weakens the critical power of the scientific method, it represents an increased leniency with respect to questions of test. Such a procedure, he says, 'propagates the acceptance of unsatisfactory hypotheses on the grounds that this is what everybody is doing. It is conformism covered up with high-sounding language' ([1962a], p. 61). In the next chapter we shall see how, and why, mid-period Feyerabend insists that there *is* a single acceptable scientific method. Relativism was a cap he was forced, by others, to wear long before he had really grown into it.[13]

It is not clear whether Feyerabend's suggested criteria of theory-comparison are supposed to be sketches of measures actually in use, or just recommendations as to how scientists *might* compare theories. Feyerabend himself was not interested in this issue. We have seen that some of his suggestions, such as formal criteria of theory-comparison, are used by scientists. Others, like syntactical measures of confirmation, at least seem to make sense, even if they play no significant role in scientific practice. But the rest of his suggested criteria seem ill-motivated, or impossible to apply.

Of course, we must remind ourselves that incommensurable theory-pairs are comparatively rare events in the history of science, despite their revolutionary role (see *FTR*, p. 272). In comparing commensurable theories, Feyerabend can also appeal to the semantic criteria which his opponents claim are the only ways rationally to compare theories. (This is not to say that these semantic criteria are unproblematic. Principled objections have been raised to criteria such as verisimilitude and/or approximate truth. It is notable both that no usable measures of these have yet been specified, and that it has not yet been shown that scientists really do use these criteria.) More work by Feyerabendians is needed to persuade us that incommensurable theories can be compared in ways which make sense and which scientists themselves deploy. But the critics' initial verdict that Feyerabend failed to indicate how incommensurable theories might be rationally compared is unjustified.[14]

We have seen that the problems surrounding Feyerabend's version of the incommensurability thesis fall into several areas:

- the vagueness of its central terms ('primitive descriptive terms', 'basic principles', 'fundamental rules', etc.);
- inadequacies in Feyerabend's extended version of the contextual theory of meaning;
- the fact that no clear concept of conceptual incompatibility has yet been forged; and
- the fact that more work needs to be done in demonstrating that incommensurable theories can be, and are, rationally compared by scientists in ways that make sense.

The incommensurability thesis produces strongly partisan reactions. Feyerabend gave various accounts of incommensurability, some of them incompatible, and some indefensible. A charitable verdict would be that his critics have not yet shown that its consequences are sufficiently absurd to justify its rejection.

7

Theoretical Pluralism

7.1   The Orthodox Test Model and the Autonomy Principle

Feyerabend's crusade on behalf of theoretical pluralism was inspired by his study of the quantum theory. While conceding that the theory had accumulated impressive empirical success, he also thought (as we have noted) that it had been designed to satisfy some over-restrictive empiricist requirements. Certain defenders of the theory in its Copenhagen Interpretation claimed that its success made it both indispensable and unassailable. Feyerabend rightly identified this as an unjustifiable claim: the fact that a particular theory is empirically adequate cannot provide grounds for believing that it will remain the only possible adequate theory. Because a given set of observation-statements can be explained by many different, incompatible theories, there is no valid argument from explanatory success to exclusive adequacy. When defenders of the Copenhagen Interpretation attacked the rival 'hidden-variable' theory, suggested by Louis de Broglie and developed by David Bohm, Feyerabend responded by arguing for theoretical pluralism. He wanted to provide room for hidden-variable theorists to pursue their ideas unhindered by arguments purporting to show that the Copenhagen Interpretation was the only reasonable point of view in microphysics.

The reason why the myth predicament does not often occur in scientific practice, according to Feyerabend, is that the test procedure followed in science is not generally governed by what he calls 'the orthodox test model', that is, the account of theory-testing implicit in positivist and post-logical empiricist philosophy of science alike. In this chapter I want to evaluate critically Feyerabend's argument for theoretical pluralism, which seeks to impugn the rival conception of theory-testing embodied in the orthodox test model. The argument purports to show that that test model fails to take account of the way in which the empirical content of a theory depends not merely on its own relations to the facts of observation, but also on the state of rival theories in the same domain.

According to the orthodox test model, a theory is a set of meaningful and truth-valued sentences. From the laws of the theory, together with statements of initial conditions, we derive a set of testable observation-statements. These predictions are then compared with the results of experiment and observation. If the two match up (within an acceptable margin of error), we regard the theory as confirmed, or at least 'corroborated', by the evidence. If the predictions fail to match the observed results, we are obliged to alter our theory, or to regard it as refuted, or to disallow the experimental result on some procedural grounds (inaccurate measurement, faulty observation, stray report, etc.). The theory's empirical content, a measure of how much information it conveys, is given by the number of its 'potential falsifiers': the testable singular observation-statements which would, if confirmed, falsify it.

Feyerabend aims to show the untenability of this apparently innocuous account of theory-testing. His fundamental contention is that it grounds the conceptual conservatism he identifies in the accounts of science developed by classical empiricists, the Logical Positivists, the logical empiricists, and Kuhn.[1] Its essential feature is its *theoretical monism*: it deals with the relationship between a single theory and a fixed set of observation-statements or facts. Feyerabend claims that this test model fails both of his criteria of adequacy for test methodologies. First, it is descriptively inaccurate to the history of science: it fails to take account of the *essential* role of alternative theories in the process of falsification. Thus it is an inadequate instrument with which to approach the history of science, or of thought in general. Second, its implementation, which would be possible, would nevertheless be undesirable, since it would facilitate, if not ensure, the development of a myth predicament, in which knowledge 'grows' only in an impoverished sense (a worthless accumulation of superficial Baconian 'facts').

Feyerabend's most powerful argument suggests that the necessity for a plurality of theories in any given domain at any time derives from considerations of testability. According to him, the orthodox test model makes a certain essential presupposition which is unexamined, and false. Surprisingly (and, I think, unwisely) he concedes that if this submerged assumption were sound there would be no good argument for theoretical pluralism. The assumption in question, the 'autonomy principle', is that the facts to be collected for the purposes of testing a theory exist, and are available whether or not one considers alternatives to that theory. This is glossed as follows:

> It is not asserted by this principle that the discovery and description of facts is independent of *all* theorising. But it *is* asserted that the facts which belong to the empirical content of some theory are available whether or not one considers alternatives to *this* theory.

Facts and theories, retorts Feyerabend,

> are much more intimately connected than is admitted by the autonomy principle. Not only is the description of every single fact dependent on *some* theory (which may, of course, be very different from the theory to be tested). There exist also facts which cannot be unearthed except with the help of alternatives to the theory to be tested, and which become unavailable as soon as such alternatives are excluded. ([1963a], p. 22; [1965a], pp. 174–5)

This is the rationale for the use of Feyerabend's 'pluralistic test model': certain facts, which may bear negatively on a proposed theory, can only be discovered by developing other theories. To attack the autonomy principle, however, is a heroic but difficult way to make the case for theoretical pluralism. Feyerabend takes this course only because he imagines that the strongest argument for proliferation has to be couched in terms of enhancing refutational possibilities. He is concerned about how we can increase the number of potential falsifiers a theory has. That is, he justifies the pluralistic test model in terms of what he calls the 'principle of testability', a methodological demand that our theories should have the maximum possible testable or empirical content. One might feel that there are other ways of making the case for proliferation, but Feyerabend is surely right in thinking that this argument would, if sound, be very important.

7.2 The Case of the Brownian Motion

From 1962 onwards, Feyerabend repeatedly used the same example to illustrate the function of alternative theories in the discovery of refuting facts.[2] Aiming to find a situation where there exist observations sufficient for refuting a theory, but where we could never establish this on the basis of the theory and observations alone, his example, originally suggested by Bohm, concerns the phenomenon of Brownian motion. Feyerabend urged that it could be generalized to produce a schema typical of the relationship between high-level theories and the facts. We shall begin by presenting a sketch of the history of Brownian motion, highlighting the facts salient for Feyerabend's case.[3]

By the early nineteenth century, many scientists had observed tiny particles in constant but irregular motion within water droplets. In 1827, the Scottish botanist Robert Brown realized that this phenomenon could not be biological, since it persisted in water droplets trapped within rock samples many millions of years old. In subsequently showing that 'almost any kind of matter, organic or inorganic, can be broken into fine particles that exhibit the same kind of dancing motion' (Brush [1968], p. 3), Brown drew the correct conclusion that this was a physical phenomenon.

Throughout the mid-nineteenth century, attempts were made to give the phenomenon a physical explanation in terms of evaporation, air currents, heat flow, electrical effects, etc. Although these explanations were shown to be spurious, most mid-nineteenth-century physicists still thought that Brownian motion would be explained by a future theory of molecular motion, but that a detailed investigation was not worth the trouble. Interest therefore dropped off for about thirty years after Brown's publications of 1828–9. A revival of interest in the relation between heat and microscopic motion only came as the result of the development, in the 1850s, of the kinetic theory of gases, which identified heat with molecular motion. Unfortunately, Brownian motion was largely ignored even by those who developed the kinetic theory, such as Rudolf Clausius, James Clerk Maxwell and Ludwig Boltzmann. But by the 1870s many physicists were prepared to affirm that it had something to do with heat, even though they had no testable quantitative theory of the phenomenon which made it amenable to the kinetic theory. Investigations into Brownian motion conducted from the point of view of the kinetic theory all had negative results.

In the late nineteenth century a rival theory of heat known as phenomenological thermodynamics held sway. Its adherents argued that there was no need to postulate the existence of atoms and molecules when phenomenological (that is, macroscopic) laws of thermodynamics were empirically adequate. As positivists, they also went further, arguing that to postulate unobservable entities (like atoms and molecules) would introduce needless paradoxes and inconsistencies. One of the laws of the phenomenological theory, the second law of thermodynamics (sometimes known as 'Carnot's principle') is particularly relevant here. This says that in thermal equilibrium heat cannot completely turn into work. That Brownian motion might constitute a counter-example to this 'law' was first suggested by the French physicist Léon Gouy. Particles in Brownian motion could, in theory, produce work at the expense of the heat in their surrounding medium without the motion conveyed to these particles by that medium diminishing. So the second law failed to hold on the molecular level. Poincaré concurred:

> If, then, these movements never cease, or rather are reborn without ceasing, without borrowing anything from an external source of energy, what ought we to believe? To be sure, we should not renounce our belief in the conservation of energy, but we see under our eyes now motion transformed into heat by friction, now heat changed inversely into motion, and that without loss since the movement lasts forever. This is contrary to the principle of Carnot. (Poincaré, quoted in Brush [1968], p. 13)

Poincaré did not, however, work out the missing quantitative theory of Brownian motion; that task was left to Einstein.

Einstein made the decisive breakthrough while trying to convince physicists of the viability of the kinetic theory. He introduced a probabilistic account of particle diffusion derived from the statistical mechanics of Boltzmann and Gibbs. Einstein later recorded that, in developing statistical mechanics and the kinetic theory, he was aiming 'to find facts which would guarantee as much as possible the existence of atoms of definite size'. 'In the midst of this', he continued,

> I discovered that, according to atomistic theory [i.e., the kinetic theory], there would have to be a movement of suspended microscopic particles open to observation, without knowing that observations concerning the Brownian motion were already long familiar. (Einstein in Schilpp [1949], p. 47)

In his investigations, conducted around 1905,[4] Einstein boldly combined two existing theories (of hydrodynamics and of osmotic pressure) which apparently applied only in different domains of validity, thereby ignoring domain distinctions (like that between the observable and the unobservable) insisted upon by the phenomenological theory. He and the Polish physicist Marian von Smoluchowski independently produced the first experimentally testable quantitative predictions applying to Brownian motion. The kinetic theory explained that it was the result of the relatively massive particles being bombarded by the ceaselessly moving molecules of the surrounding fluid.

The French physicist Jean Perrin is generally credited with having experimentally established the Einstein–Smoluchowski theory, and with thus having confirmed the kinetic theory. Perrin, following the work of Theodor Svedberg and others, conducted a series of experiments designed to test a simplified version of Einstein's theory. He too was convinced that Brownian motion was a counter-example to the second law of phenomenological thermodynamics. His confirmation of Einstein's predictions about diffusion supported Einstein's theory and demonstrated that the second law had to be fundamentally revised (even though it could still be regarded as 'statistically valid'). Perrin's work led to an experimental determination of Avogadro's number (the number of atoms in one mole of a substance), the mass and size of an atom, and the charge on the electron. The ultimate result of his experiments was acceptance, by physicists, of the physical reality of atoms. Even some of the most hardened positivists came to believe in them by the end of the first decade of the twentieth century. The only major dissenter was the instrumentalist Ernst Mach, who persisted in treating the existence of atoms as a fiction, a useful hypothesis. Brownian motion is nowadays regarded as the best illustration of the existence of random molecular motions, a 'conclusive observational method for confirming the atomic theory of matter' (Lavenda [1985], p. 56).

Now let us consider Feyerabend's claim.[5] He does not deny that the relevance of Brownian motion could have been discovered merely by considering the phenomenon and the phenomenological theory. (We have seen that its relevance was thus recognized by Gouy and Poincaré.) However, he contends that the fact that the Brownian particle, seen from a microscopic point of view, is a *perpetuum mobile* of the second kind, and thus refutes the second law,[6] could not have been discovered by an investigation of the observational consequences of the phenomenological theory alone, without borrowing from an alternative account of heat. A 'direct refutation' of the theory, a refutation involving that theory and this phenomenon alone, is, he says, 'excluded by physical laws' ([1966a]: *PP2*, p. 62). To discover that the phenomenon and the theory are inconsistent with each other would have required:

(a) measurement of the exact *motion* of the particle in order to ascertain the change in its kinetic energy plus the energy spent on overcoming the resistance of the fluid; and (b) it would have required precise measure-

ments of temperature and heat transfer in the surrounding medium in
order to establish that any loss occurring there was indeed compensated
by the increase in the energy of the moving particle and the work done
against the fluid. Such measurements are beyond experimental possibilities.... Hence a 'direct' refutation of the second law that would consider
only the phenomenological theory and the 'facts' of the Brownian motion, is impossible. It is impossible because of the structure of the world
in which we live and because of the laws that are valid in this world. (AM^1,
p. 40)

Being concerned only with the relation between the theory under test
and the evidence, the orthodox test model cannot take into consideration the state of any rival theory. A 'direct' refutation is therefore the
only sort it can comprehend. The verdict that a direct refutation was
impossible in the Brownian motion case is also suggested by the fact
that the actual refutation of the phenomenological theory was not of
this sort: in the actual refutation Einstein relied upon the kinetic
theory to explain both the deviations of the Brownian particle from
the second law and certain new facts. This new explanation was
judged superior to the old theory in part at least because it fared
better in the 'crucial experiment' staged by Perrin. Thus the phenomenological theory was overcome in what Feyerabend calls an
'*indirect refutation*'.

7.3   The Generalized Refutation Schema

The case of the Brownian motion is not the only example of an indirect
refutation Feyerabend offers although it is, unfortunately, the only
one he develops in any detail.[7] It is alarming, therefore, when he
simply conjectures that this scenario 'is *typical* of the relation between
fairly general theories, or points of view, and the "facts"' (AM^1, p. 41,
emphasis added).[8] In more considered presentations, he does not
make this hasty transition, but elaborates what I (following Ronald
Laymon) shall call the Generalized Refutation Schema, which I here
quote in full:

> We may generalise this example as follows: assume that a theory T
> has a consequence C and that the actual state of affairs in the world
> is correctly described by C', where C and C' are experimentally indis-
> tinguishable. Assume furthermore that C', but not C, triggers, or causes,
> a macroscopic process M that can be observed very easily and is per-
> haps well known. In this case there exist observations, viz., the observa-
> tions of M, which are sufficient for refuting T, although there is no
> possibility whatever to find this out on the basis of T and of obser-

vations alone. What is needed to discover the limitations of T implied by the existence of M is another theory, T', which implies C', connects C' with M, can be independently confirmed, and promises to be a satisfactory substitute for T where this theory can still be said to be correct. Such a theory will have to be inconsistent with T, and it will have to be introduced not because T has been found to be in need of revision, but in order to discover whether T *is* in need of revision. ([1965a], p. 176)

The Generalized Refutation Schema represents Feyerabend's attempted refutation of inductivism, specifically, of the 'modified generalization hypothesis', promised in chapter 1. That hypothesis, recall, said that given a finite collection of singular statements Pa, Pb, . . . Pn (abbreviated 'P(n)'), it is reasonable to infer (x)Px. Feyerabend disagreed, or at least he disagreed if the hypothesis is taken to mean that this inference is the only reasonable one that can be inferred from those data. He committed himself to refuting the hypothesis, and thereby showing that the procedure it supposedly suggests, to restrict attention exclusively to the 'theory' (x)Px, is undesirable. To do so, he argued that the existence of cases where there are hidden refuting facts cannot be ascertained in advance:

> *Any* theory T under consideration . . . may be inconsistent with facts which are accessible only indirectly, with the help of an alternative T'. Now it is surely reasonable to demand that the class of refuting instances of a given theory be made as large as possible and that especially those facts which belong to the empirical content of the theory, which refute the theory but which cannot be distinguished from similar, but confirming facts, be separated from the latter and be thus made visible. However, this means that, given P(n), it is reasonable to use not only (x)Px, *but as many alternatives as possible*. This is the promised refutation of the modified generalisation hypothesis. ([1964b]: *PP1*, p. 206)

This underlines the pivotal role of the Generalized Refutation Schema in Feyerabend's positive philosophy. Only if that schema works has he shown that instrumentalism, theoretical monism, cumulativism, inductive methodology and the orthodox test model are unacceptable to a tolerant empiricist.

This aspect of Feyerabend's philosophy has recently come in for detailed critical examination. Ronald Laymon, John Worrall and Larry Laudan have all cast doubt on Feyerabend's argument, although not in the same way. Their basic contention is that on any ordinary understanding of the notion of empirical content, the Generalized Refutation Schema does not work.

7.4   A New Conception of Empirical Content?

Feyerabend's crucial claim is that there exist refuting instances of a theory T which can only be identified through developing alternatives to T. There are two incompatible ways of interpreting this.

Laudan points out that it may only be the epistemic claim that alternative theories enable us to discern that certain things already entailed by T are false. Refutation of T is then supposed to proceed as follows: we observe M; we know that it is caused by C' and not by C; the occurrence of C' confirms the alternative theory T', and hence the observation of M can be said to confirm T'. Since T entails C, and since C' and C are mutually inconsistent, we know that C did not take place: hence T is refuted.[9]

Laudan complains that there is a problem here: we only have reason to believe that M is caused by C' if we *already* accept T'. We are not told why this should persuade anyone who instead accepts T that it is now refuted. Adherents of T can surely stick to their theory and, since T' is essential for deriving M, deny that M is a potential falsifier of T. Putting this in terms of the Brownian motion example, the problem is that no defender of the phenomenological theory of heat is going to be compelled to accept that Brownian motion is caused by the movement of molecules: they do not believe in the existence of molecules.

This is true. But although defenders of the phenomenological theory may insist that postulating molecules is unnecessary for the explanation of Brownian motion, the success of the rival kinetic theory in explaining the phenomenon (whose reality everyone concedes) should give them pause for thought. They, after all, have *no* satisfactory explanation of Brownian motion. The kinetic theory, if true, would explain the observed phenomenon, M. The inference from M to the theory T' is therefore a straightforward case of inference to the best explanation. Short of coming up with an alternative explanation of Brownian motion, the phenomenological theorist must accept that the phenomenon now constitutes evidence in favour of the kinetic theory (although it seems a bit much to describe this as a refutation of the phenomenological theory). I therefore think that, on this modest interpretation, Feyerabend's argument can be defended against Laudan's charge, unless Laudan can persuade us to reject all forms of scientific realism by renouncing inference to the best explanation completely.

The other possible interpretation has Feyerabend making the more radical semantic claim that we can tell that certain observation-

statements are logically inconsistent with a theory only by considering alternative theories. Against this interpretation, Laudan reminds
us that the empirical content of a theory is defined as the class of its
potential falsifiers, and potential falsifiers are negations of the empirical consequences of the theory. The radical semantic claim is thus
unintelligible, since Feyerabend would be committed to saying that
'we can ascertain what a statement (viz., a theory) *entails* only by
considering its contraries' (Laudan [1989a], p. 310). This view, as
Laudan puts it, rides roughshod over what is commonly meant by
'empirical content' and 'entailment'. To claim that the entailments of
a theory can alter without any alteration in the theory itself is
implausible: the existence of logical relations between propositions
does not depend on whether people are capable of recognizing such
relations.

Unfortunately, there is plenty of evidence that Feyerabend did
mean to psychologize the notion of empirical content in this way. His
pluralistic test model is a development of Popperian themes. The
ultimate motivation for the 'principle of testability', which is the
single most important methodological principle behind that test
model, lies in Feyerabend's conception of empiricism (outlined in
chapter 5). He offers us a 'persuasive definition' of empiricism,
inspired by a conception popular among the Logical Positivists in the
1930s, and according to which empiricism in its most reasonable form
is a normative epistemology which 'suggests an *ideal* which we use to
criticise or praise theories, statements, points of view which are
announced together with the claim that they represent knowledge'
([1963c], p. 319). Thus empiricism is a suggestion about what the
content of our knowledge should be, not about what it is: it is the
demand that our theories and hypotheses should have the highest
attainable degree of empirical content.

At this point, recall the equation of empirical content with testability: 'a theory has factual content [if and] only if it is testable' ([1961d],
p. 402). This equation is common to Carnap, Popper and Feyerabend,
and is in each case central to their philosophies. But, true to form,
Feyerabend gives it a meaning slightly different from those which
Carnap and Popper give to it. For Popper a theory is highly testable
if its defenders have specified a large class of 'potential falsifiers',
where these are 'basic statements': testable singular existential statements referring to observable macrophysical events. For Feyerabend
too, testability, and thus empirical content, is measured by the number
of potential falsifiers a theory has. But potential falsifiers need not be
Popperian 'basic statements'. The lesson I think Feyerabend wants to

draw from the history of science (in particular, from the 'myth predicament') is that the real potential falsifiers of scientific theories are *other* scientific theories (or, at least, empirical statements which have integral connections with other theories). Given the 'basic principle of empiricism', the demand to increase the empirical content of our theories, and the equation of empirical content with testability, the use of a test model in which theories are proliferated is itself demanded by empiricism. So Feyerabend argues.

In a sense, this is just to spell out what was implicit in Popper's doctrine all along. Popper might be said to accept (even to originate) the idea that theories are falsified only by other theories. For him, a theory is falsified when we have established

> a *reproducible effect* which refutes the theory. In other words, we only accept the falsification if a low-level empirical hypothesis which describes such an effect is proposed and corroborated. This kind of hypothesis may be called a *falsifying hypothesis*. (Popper [1959], pp. 86–7)

Whether we take the premises of a falsifying argument to be low-level empirical hypotheses, or 'basic statements' themselves, Popper is committed to the view that only theories can falsify other theories, since he sees *any* such statements as theoretical ([1959], p. 75).[10] Because of the lack of an epistemologically or semantically relevant distinction between observation-statements and theoretical statements, potential falsifiers can, officially, be theoretical to any degree whatever. What Feyerabend adds to Popper's view is the idea that low-level 'theories' can only perform the function of refutation to the extent that they are informed by high-level theories.

In Feyerabend's argument for theoretical pluralism it is important to recognize, once again, the influence of the contextual theory of meaning:

> [J]ust as the meaning of a term is not an intrinsic property of it, but is dependent upon the way in which the term has been incorporated into a theory, *in the same manner the content of a whole theory* (and thereby again the meaning of the descriptive terms which it contains) *depends upon the way in which it is incorporated into both the set of its empirical consequences and the set of all the alternatives which are being discussed at a given time.* Once the contextual theory of meaning has been adopted, there is no reason to confine its application to a single theory, or a single language, especially as the boundaries of such a theory or of such a language are almost never well defined. ([1962a]: *PP*1, p. 74; emphasis added)

The contextual theory provides the motivation for the new conception of empirical content Feyerabend wants to introduce, in which potential falsifiers are alternative theories. In the orthodox test model, where a single theory T is compared with the facts, T can be rendered unfalsifiable and therefore metaphysical. In the new, pluralistic test model, each theory is highly falsifiable, maximally testable, maximally scientific, and thus minimally metaphysical, because of the actual presence of the alternative theories. Those who criticize Feyerabend by saying that his conclusions are impossible given that empirical content is as (for example) Popper says, are missing the point. Although Feyerabend started off working with the usual conception of empirical content, the onset of the pluralistic test model caused him to work out a new conception, one which makes contact with historical studies of science *and* with our intuitions about the conditions under which theories are genuinely informative. Certainly, Feyerabend never pauses to elaborate this conception or to consider its merits and defects. Nevertheless this, and not the more popular Popperian conception, is the one which he is working with:

> [T]he empirical content of a theory of the generality of the present quantum theory depends to a decisive degree on the number of alternative theories which, although in agreement with all the relevant facts, are yet inconsistent with the theory in question. The smaller this number, the smaller the empirical content of the theory. ([1962b], p. 256. See also [1963e], p. 96)

This new conception of empirical content explains Feyerabend's horror of the myth predicament, the scenario in which a theory which has no rivals becomes totally unfalsifiable, and gives every appearance of being an 'absolute truth'. It also dovetails with his pluralistic methodology.

Before we look at this methodology, there is a question to be asked about Feyerabend's conception of testability. Does it make sense to demand that we maximize the testability of our theories (or even, as Feyerabend sometimes requests, of 'our knowledge')? Testability does not seem to be a virtue of scientific theories which *can* sensibly be maximized. Each empirical claim can be said to have some degree of testability. But not every statement in a theory can be treated as testable at the same time. Some such statements function not as empirical claims but as *rules*. While Feyerabend officially recognizes this, and sometimes seems to recognize that it makes no sense to ask whether rules are true or false, he persists in demanding that we treat rules as testable, that is, as factual claims.

It is also evident that neither Feyerabend's argument for theoretical pluralism nor his argument for theoretical realism will survive the withdrawal of his normative conception of methodology. The driving force behind these methodological arguments is the unrestricted use of supposed methodological principles. Unless they are willing to let these principles override any other considerations, Feyerabend has not shown that scientists have any reason for being theoretical pluralists or scientific realists. Unless these are methodological rules which scientists actually subscribe to, they are just philosophical propaganda.

7.5 Pluralistic Methodology

According to Feyerabend, the orthodox test model fails to take full account of the fact that observation-statements are corrigible and can be reinterpreted in the light of new theories. Using this test model restricts the possibilities for refuting and improving our cherished theory, since it fails to recognize that it is too easy to reinterpret recalcitrant observations and observation-statements to make them conform to such a theory. We might never achieve a decisive refutation of our theory in this way: such a relatively feeble test procedure is conducive to the protracted survival and dominance of flawed hypotheses. Nagel's layer-cake model, for example, advises that new theories should agree, in certain very fundamental respects, with their predecessors. If we follow this advice we shall have a series of theories all of which retain the fundamental structural assumptions, both explicit and tacit, of our earliest gropings. These assumptions will be, in a real sense, unchallenged: they will constitute a privileged baseline against which new suggestions will be measured. They will have become, to all intents and purposes, a methodological *a priori*, firmly and finally delimiting the bounds of sense. An empiricist who rejects the factual/conceptual distinction cannot rest content with such a state of affairs.

For one theory thus to opt out of what Feyerabend calls the 'ocean of alternatives' in an attempt to become a foundation for all acceptable theories is for it to drop out of the domain of knowledge altogether. Of course, if scientists do not take the methodologies suggested by philosophers of science seriously, this problem may not arise. But Feyerabend felt that if philosophy of science is to be more than a descriptive catalogue of successive mini-methodologies, and if it is worth taking seriously, there must exist a methodology which tallies with the consistently revolutionary aspects of scientific change; in short, a pluralistic methodology.[11]

Again, consider the case of Kuhn. His one-sided reading of his own model of scientific development encourages the proponent of a mature science not seriously to consider alternative explanations of the phenomena that are being dealt with. It is difficult to see how there could ever be a scientific revolution if we follow this advice. For Feyerabend, Kuhn's model of science, read in the right way, is acceptable (if unsophisticated) and supports theoretical pluralism. But it is overlaid with a wholly unacceptable monistic ideology which, if given free rein, could have stultifying effects upon the growth of knowledge.

What will a pluralistic methodology look like? Feyerabend conceives of it as a generalization of the idea of a crucial experiment. The pluralistic test model is a way of assessing the merits of rival theories even when those theories are incommensurable, and thus cannot be compared (using the more familiar semantic criteria) in an ordinary crucial experiment. It will be composed of certain principles or 'demands', high-level methodological rules not just for deciding between existing theories, but also for regulating the construction of new theories. The principles Feyerabend proposes during the relevant period, before the onset of 'epistemological anarchism',[12] are as follows:

- PRINCIPLE OF FALSIFICATION Take refutations seriously ([1958a]: *PP1*, p. 34; [1958c]: *PP1*, p. 236; [1960c]: *PP1*, pp. 233, 235; [1969c], p. 253).
- PRINCIPLE OF REVISION Admit no unrevisable statement into the body of our knowledge. In any conceptual change, regard no statement, whether 'theoretical' or 'observational', as incorrigible, irrefutable, unalterable, or *a priori* ([1958a]: *PP1*, p. 35; [1960a], p. 64; [1962a], pp. 39–40; [1965a], p. 184; [1965c]: *PP1*, p. 125).
- PRINCIPLE OF EMPIRICISM Maximize the empirical content of existing theories ([1958a]: *PP1*, p. 35; [1962a]: *PP1*, pp. 60, 72; [1963a], p. 29; [1964b]: *PP1*, p. 206; [1965a], pp. 176, 196; *AM*[1], p. 41).
- PRINCIPLE OF TESTABILITY Use only theories which are testable. Put theories in a form in which they are maximally testable, and test them (attempt to falsify them) relentlessly ([1960c]: *PP1*, p. 235; [1962a]: *PP1*, p. 45; [1963c], p. 322; [1964a]: *PP1*, pp. 200–1; [1965a], p. 196; [1965b]: *PP1*, p. 102; [1965c]: *PP1*, p. 105).
- PRINCIPLE OF REALISM Develop theories in their strongest possible form, i.e. as attempts at universally quantified descriptions of reality rather than as mere instruments of successful prediction ([1958a]: *PP1*, p. 34; [1962a]: *PP1*, p. 45; [1964a]: *PP1*, p. 200; [1966a]: *PP2*, p. 63).[13]

- PRINCIPLE OF PROLIFERATION   Invent, and elaborate, theories which
  are inconsistent with the accepted point of view, even if the latter
  should happen to be highly confirmed and generally accepted
  ([1965c]: *PP1*, p. 105; [1968c], p. 280; [1970a]: *PP2*, pp. 139, 143–4).[14]
- PRINCIPLE OF TENACITY   From a number of theories, select the one
  that has the most attractive features and that promises to lead to the
  most fruitful results. Stick to it even if it is inconsistent with
  evidence, or encounters considerable difficulties ([1968b], p. 131;
  [1968c], p. 280; [1970a]: *PP2*, pp. 137–8, 143–4).

There is no indication that Feyerabend ever put these rules together
in his mind as a single pluralistic methodology; nevertheless they are
repeatedly referred to in his work over a significant period of time. If
we take this pluralistic methodology seriously, we will work, not
with a single theory, however plausible, but with '*a whole set of partly
overlapping, factually adequate, but mutually inconsistent theories*' ([1962a]:
PP1, p. 72). This supposedly generates a new conception of knowledge itself:

> Knowledge so conceived is not a process that converges towards an ideal
> view; it is an ever-increasing ocean of alternatives, each of them forcing the
> others into greater articulation, all of them contributing, via this process
> of competition, to the development of our mental faculties. ([1965c]: *PP1*,
> p. 107)

(Whether this makes sense as a conception of *knowledge* is, of course,
a moot point.) The pluralistic test model demands that theories be
interpreted in their strongest form (realistically), that they be strongly
inconsistent with each other, that the inconsistency is not just on the
more highly theoretical levels, and that we do not abandon but
develop and elaborate any theories which we regard as being refuted.
It is not an alien God to which scientists must be forced to pay
homage, but is already in our midst, and has been at work in the best
science since the pre-Socratics. It is simultaneously a better description of what happens in science, and a more worthy ideal, than those
offered by more 'monistic' thinkers.

We might note the irony of Feyerabend's appealing so fervently to
principles of methodology when he thinks, initially, that there are no
facts of the matter about methodology and, latterly, that there is no
such thing as scientific method. By arguing that it is wrong even to
think that philosophers can lay down (rather than merely uncover)
methodological rules, I have already tried to cut off the reply that
these rules are determined by the philosopher. (Of course, philoso-

phers might suggest methodological rules to scientists. But whether there is a scientific method depends on whether scientists do follow rules).

Theoretical pluralism (that is, Feyerabend's pluralistic methodology) should not be confused with *methodological* pluralism. Commentators sometimes take Feyerabend to be advocating the latter even in his pre-1970 work, but this is a serious mistake.[15] The pluralistic test model is intended to be a single methodology for all scientific inquiry. It sponsors the proliferation of theories, but not of methods for evaluating theories. Only after the advent of 'epistemological anarchism' does Feyerabend propose that science has no distinctive method, and that it cannot be forced by philosophers to accept one. When Feyerabend became a methodological pluralist, he had officially forsworn the resources which originally allowed him to argue for theoretical pluralism.

7.6 Which Principle of Proliferation?

Feyerabend's pluralist crusade has sometimes been accused of being a war against straw Turks. We saw that this is not true: there are theoretical monists and cumulativists. But Feyerabend claims that the pluralistic test model goes hand in hand with the 'principle of proliferation', which he encapsulates in the slogan 'Let a thousand flowers bloom'. Formulations of (rather than mere appeals to) this methodological principle are rare in Feyerabend's work. In perhaps his most considered version it exhorts us to *'invent, and elaborate theories which are inconsistent with the accepted point of view, even if the latter should happen to be highly confirmed and generally accepted'* ([1965c]: *PP1*, p. 105). Although the principle occurs over a significant period in Feyerabend's work, his understanding of it changes. Initially, he understands it to require that newly introduced theories should have various virtues (empirical adequacy, etc.). But he slides from this more modest principle to what I shall call an unrestricted principle of proliferation which puts no restriction whatsoever on the calibre of the alternative theories. It invites us to invent, elaborate and retain theories no matter how inadequate they are. It counsels not only the invention of new theories but also the retention of older theories which have been refuted. Is the pluralistic test model wedded to this version of the principle?

It is not. Even if Feyerabend's argument for theoretical pluralism succeeds, the unrestricted principle of proliferation still does not follow. Feyerabend's thought is: the more theories we invent, the higher the empirical content of all, however weedy the invented newcomers are. This simply does not follow either from the Brownian motion example or from the Generalized Refutation Schema: in both those cases an 'indirect refutation' is achieved only because the alternative theory deployed is at least as adequate as the existing theory. The examples just do not begin to work unless the new theory reproduces the success of the old one and makes novel predictions too, so only an adequate new theory can 'indirectly refute' an old one. The pluralistic test model works with 'a whole set of partly overlapping, *factually adequate*, but mutually inconsistent theories', not any old bunch of half-baked ideas. This 'condition of factual adequacy' cannot, *pace* Feyerabend, be 'removed' (AM^1, p. 41, n. 8), for it is only an acceptable theory which has a chance of truly superseding ('falsifying') and thereby replacing an older theory. The transition to the unrestricted principle of proliferation involves a fallacy of subtraction. *Good* alternative theories are 'necessary parts of the falsifying process'!

None of this means that it is fruitless to attempt to develop half-baked ideas into good theories. It is just that there is a threshold point, below which at any given time the theories which exist will be incapable of affecting the empirical content of acceptable theories. (Consider the parallel case of measuring instruments: not just any new instrument will be capable of facilitating new tests of existing theories.)

Feyerabend's *a priori* methodological argument for scientific realism (which we left back in chapter 4) is now revealed as ultimately untenable. The argument was to the effect that only a realist could support the generous principle of proliferation required to maximize the efficiency of the pluralistic test model. But that model does not work with the raw material which Feyerabend thinks it does. To expose refuting facts an adequate theory, a rival which can be taken seriously, is necessary. With the collapse of the central argument for the unrestricted principle of proliferation, Feyerabend's argument for realism also suffers. The tortuous path through pluralism to realism is, as far as I can see, one which never reaches its destination.

His long *a priori* argument from proliferation is also ultimately irrelevant since, as Giedymin has shown, anti-realists can subscribe to equally strong principles of proliferation. We might even suspect that the principle of proliferation would work in the opposite direction: it is the instrumentalist, after all, who retains theories and feels no hesitation in using them when the realist would renounce them as falsified. If any and every theory makes a contribution to the empirical content of each, the realist's insistence on using only the most developed theories is an obstacle to theoretical pluralism, not a guarantor of it.

8

Materialism

8.1 Super-Realism

Twentieth-century philosophy offers several accounts of the relationship between science and common sense. A traditional way of introducing the issue is by considering Eddington's two tables. Arthur Eddington, a physicist who led one of the expeditions to conduct the crucial experiment intended to confirm Einstein's theory of relativity, began his book *The Nature of the Physical World* by asking us to consider the two tables at which he was writing:

> One of them has been familiar to me from earliest years. It is a commonplace object of that environment which I call the world. How shall I describe it? It has extension; it is comparatively permanent; it is coloured; above all it is *substantial*. . . . Table No. 2 is my scientific table. It is a more recent acquaintance and I do not feel so familiar with it. It does not belong to the world previously mentioned – that world which spontaneously appears around me when I open my eyes . . . It is part of a world which in more devious ways has forced itself on my attention. My scientific table is mostly emptiness. Sparsely scattered in that emptiness are numerous electric charges rushing about with great speed; but their combined bulk amounts to less than a billionth of the bulk of the table itself. (Eddington [1928], pp. xi–xii)

The question of the relationship between science and common sense is sometimes put in terms of the objects they postulate: since there are *not* two tables in the offing, which of Eddington's 'two' tables is the (one and only) real one? Another way of putting the question is in terms of Wilfrid Sellars' distinction between the 'manifest' and 'scientific' images of the world. The former is the world as represented in terms of the common-sense conceptual framework, the latter the world as seen by theoretical science. The question then becomes: which of these two images is the more correct? The main answers on offer are roughly as follows:

1   Strict instrumentalists, if they existed, would respond that the ordinary table is real, and that 'the scientific table' is a fiction generated by taking too literally statements of scientific theory which should really be understood as rules for predicting the occurrence of perceptions. Thus there is no conflict between science and common sense since only observation, and not theory, can tell us what is real.

2   Philosophers like Ernest Nagel, for whom the dispute between realism and instrumentalism is 'a conflict over preferred modes of speech', answer that the question contains an error, a failure to recognize that statements in the language of theoretical physics cannot be equivalent in meaning to statements of 'ordinary language':

> Eddington was not at all confronted with two tables. For the word 'table' signifies an experimental idea that does not occur in the language of electron theory; and the word 'electron' signifies a theoretical notion that is not defined in the language employed in formulating observations and experiments. Though the two languages may be co-ordinated at certain junctures, they are not intertranslatable. Since there is thus only one *table*, there is no issue as to which is the 'real' one, whatever may be understood by this honorific label. (Nagel [1961], pp. 98–9n)

For Nagel, common sense is both indispensable and well on the way to truth; so is science. They do not conflict, since they have no common subject matter.

3   Another compatibilist view, reductionism, would claim that statements of common sense are deducible from those of theoretical science. For reductionists, both science and common sense are acceptable, and they are compatible with one another.

4   Moderate scientific realists like Sellars and J. J. C. Smart sometimes take the view that common sense is false, but still indispensable. On this view, the scientific image of nature is, or will eventually be, the true image. Nevertheless, we cannot simply dispense with the manifest image, since that image is

> the conceptual structure in which we naturally think, and in which we cannot help thinking so long as we do not deliberately and consciously determine to think in terms of scientific theories. . . . According to Sellars . . ., the manifest image is quite indispensable. It is only through the

manifest image that we can get to theoretical science at all. (Smart [1965],
p. 166)

As a result, we must operate with two different and inconsistent
images of the world. *Pace* reductionism, the statements of the manifest image are not deducible from those of the scientific image.

5   Finally, the most extreme scientistic view is that science and
common sense are in flat-out conflict, and that whereas science is our
best guess as to the truth, common sense is neither indispensable nor
true. Common-sense theories may be useful for practical purposes, but
their theoretical credentials are flawed. We therefore ought to think of
the world in scientific terms, rather than those of common sense. The
result of accepting physical theory is that our everyday beliefs about
the world must be totally replaced. This is Feyerabend's view, sometimes known as 'super-realism'. Despite being a convinced materialist,
Feyerabend arrived, via super-realism, at the conclusion that there are
no material objects as traditionally conceived.

Scientific theories often tell us of the existence of new and unsus-
pected things, but they do not pronounce on what is real, or on what
alone is real. Super-realism is far less modest: it goes beyond scientific
realism in claiming that only what our best theories postulate is real.
How did Feyerabend arrive at such conclusions? We already know
that his scientific realism asks us to interpret observation-statements
in terms of the theories we accept. According to Feyerabend, a
successful high-level theory should be capable of being its *own*
observation-language. A Feyerabendian realist 'wants to give a unified account, both of observable and of unobservable matters, and he
will use the most abstract terms of whatever theory he is contemplat-
ing for that purpose' ([1970a]: *PP2*, p. 153). Such a theory will supply
us with theory-laden observation-statements in terms of which we
can directly conceive the nature of observable things. I have agreed
with Feyerabend that scientific observation-statements will usually
be statements about material objects, not sense-datum statements.
There then follows a tempting line of argument.

The material-object statements which serve as scientific observation-statements should be interpreted in terms of our most successful
fundamental theories of matter. So we should look to microphysics
for information about the real nature of material objects. Common
sense and classical physics coincide, to a great extent, in their conception of material objects. In particular, they both agree that material
objects continue to exist even while they are not being perceived.
Feyerabend, who calls this the idea of 'observer independence',

argues that it is refuted both by contemporary microphysics and by Einstein's theory of relativity:

> Einstein . . . suspected that the classical point of view was itself incorrect, and constructed a theory which was inconsistent with the idea of observer independence. In comparing this theory with the classical solution . . . it was found to possess tremendous advantages, and it was therefore adopted. . . . The consequence of all this is, of course, that material objects . . . (i.e. objects which are observer-independent and which exist unperceived) *do not exist*. And as tables, chairs, bookcases, are supposed to be material objects . . . it follows that they do not exist either. ([1969b]: *PP1*, p. 158)

Notice that this view goes beyond even that of Berkeley, who held that observer-independent material objects do not exist, but that chairs, tables, etc., *do*, since they are not material objects but collections of ideas. Feyerabend thinks that physics, interpreted 'realistically', tells us that the entire world of familiar material objects is a mere 'phenomenon', an illusion. In the place where we think the solid table stands there is nothing but a congeries of swiftly moving subatomic particles, and even they are not material objects as traditionally conceived. Since this conclusion contradicts the common-sense view, sophisticated scientific theory is inconsistent with common sense: we cannot believe both. Which shall we side with? The scientific theory represents the pinnacle of our intellectual achievements: it is precise, explanatorily and predictively powerful, and well confirmed. The common-sense view, on the other hand, may be well confirmed, but it has nothing else going for it. It is old, but age is no virtue in a theory. Given the choice, then, we must side with science. All that exist are the strange subatomic entities of microphysics and, perhaps, the void.

Feyerabend allows that the conceptual revolution he has in mind may seem strange, but insists that it is no more strange than the denial of angels, demons or the devil seemed to faithful religious people who had been brought up to believe in them, and who had had the corresponding experiences. For these people, 'the spiritual world . . . was much more important and much more secure than the transitory world of tables, chairs, and philosophy books' (*PP1*, p. 159).

8.2 Science and Material-Object Concepts

I believe that almost every component of this line of argument should be vigorously resisted. It certainly *has* been resisted, from different quarters. I hope here to indicate several lines of resistance which might be compressed into a case against super-realism.

Recall Feyerabend's admission that scientific observation-statements *generally* make reference to material objects rather than to sense-data (although there are statements very like sense-datum statements in psychology and physiology). This should be our bridge-head.

One way of developing this thought is to follow Charles Peirce in arguing that the initial subjects of inquiry are those apprehended in common sense, and that common-sense concepts pick out a relatively theory-neutral subject matter which is not affected when we reject the positivist idea of theory-independent observation. Common-sense concepts are neither inadequate nor even seriously corrigible, since they are *vague*: 'Our most primitive descriptive concepts provide a means of identifying the subjects common to rival theories, not because these concepts are phenomenalistic, but because they are explanatory as well as descriptive, only very vaguely so' (Short [1980], p. 323).[1] Common-sense concepts are still theory-laden, but the theory they are laden with is the very indeterminate one called 'common sense'. This line of argument ties in with Feigl's familiar idea (mentioned in chapter 3) that ordinary-language observation-statements are relatively theory-neutral.

Another powerful line of argument, deriving from the work of Peter Strawson, culminates in the claim that our ordinary concept of a material object has an irreplaceable role in our conceptual scheme. The reasoning here is as follows. The identification of particulars of one kind often depends on the identification of particulars of another kind. Where this identifiability-dependence is most simple and direct, particulars of the latter kind can be said to be more basic to our conceptual scheme than those of the former kind. Now in order for particular-identification to be possible, there must exist a common frame of reference, a single system of relations in which each identifiable particular has a place. The system of spatial and temporal relations, space and time themselves, constitute this frame of reference.

Strawson now asks: is there any class of particulars which *must* be basic from the point of view of particular-identification? From the premise that particular-identification rests ultimately on location in a unitary four-dimensional spatio-temporal framework, he argues to the conclusion that a certain class of particulars is indeed basic:

[T]hat framework is not something extraneous to the objects in reality of

which we speak. If we ask what constitutes the framework, we must look
to those objects themselves, or some among them. But not every category
of particular objects which we recognise is competent to constitute such a
framework. The only objects which can constitute it are those which can
confer upon it its own fundamental characteristics. That is to say, they must
be three-dimensional objects with some endurance through time. They
must also be accessible to such means of observation as we have. (Strawson
[1959], p. 39)

It is (at least) a necessary condition of basic particulars that they
should be public objects of perception, 'particular objects of such
kinds that different people can quite literally see or hear or feel by
contact or taste or smell the same objects of these kinds' ([1959], p. 45).
Material objects are the very best candidates for this status of basic
particulars. They play a unique and fundamental role in particular-
identification. The category of material objects, and that category
alone, Strawson concludes, 'supplies enduring occupiers of space
possessing sufficiently stable relations to meet, and hence to create,
the needs with which the use of [our unified spatio-temporal framework of particular-reference] confronts us' (p. 56). All things which
are or have material bodies qualify as basic particulars.

According to this view, the theoretical constructs of science are
identifiability-dependent on material objects in a direct way. The
objects which such concepts purport to pick out, like the particles of
microphysics, says Strawson,

are not in any sense private objects; but they are unobservable objects. We
must regard it as in principle possible to make identifying references to
such particulars, if not individually, at least in groups or collections;
otherwise they forfeit their status as admitted particulars. . . . But it is clear
enough that in so far as we do make identifying references to particulars of
this sort, we must ultimately identify them, or groups of them, by identifying reference to those grosser, observable bodies of which perhaps, like
Locke, we think of them as the minute, unobservable constituents. ([1959],
p. 44)

On this view, therefore, theoretical languages are parasitic on ordinary language; even the language of microphysics must piggyback
on the material-object language, and cannot, as Feyerabend envisaged, replace it. Theoretical entities may well exist, but they cannot be
basic particulars.

Feyerabend did not explicitly deal with such transcendental arguments, but he did consider certain objections to his super-realism,
such as the view that 'the language in which we describe our surroundings, chairs, tables and also the ultimate results of experiment

(pointer-readings) is fairly insensitive towards changes in the theoretical "superstructure"' ([1958a]: *PP*1, pp. 30–1). He complained,
first, that this view wrongly presupposes the existence of a uniform
'everyday language', where in reality there is only a mixture of
heterogeneous languages; and, second, that this mixture does undergo important changes:

> terms which at some time were regarded as observational elements of
> 'everyday language' (such as the term 'devil') are no longer regarded as
> such. Other terms, such as 'potential', 'velocity', etc., have been included
> in the observational part of everyday language, and many terms have
> assumed a new use. (*PP*1, p. 31)

But Feyerabend conceded that scientists, who use our ordinary
material-object language in describing their experiments, do *not*
introduce a new use for the terms they deploy in these descriptions,
like 'pointer' and 'red'. He thought, however, that this did not
establish that those words do not have a new meaning. The considerations adduced in chapter 2 indicate that his account of meaning, with
its insistence that the meaning of a word can change without its use
changing, is inadequate. Feyerabend's concession that the use of
these terms remains unchanged is the crucial point.

Feyerabend did several times consider the objection, which he
associated with positivists and logical empiricists such as Carnap,
Feigl and Nagel, that it is simply impossible to replace our everyday language with a truly theoretical language. The idea is that any
new language or conceptual framework must be related to the
everyday language which embodies our common-sense conceptual
framework, because 'unless a connection is established with previous
language, we do not know what we are talking about, and we are
therefore not able to formulate our observational results' ([1963d], p.
296). On this view, theoretical terms receive their interpretation by
being connected either with a pre-existing observation-language or
with another theory already connected with such an observation-
language. Without such a connection they are meaningless. If this is
so, if theoretical terms do not have a meaning which is independent
of their theory's observation-language, they cannot be used to update
the meaning of observation-statements (which are their only source
of meaning). Feyerabendian realism would thus be untenable.

Such an objection, says Feyerabend, 'assumes that the terms of a
general point of view and of a corresponding language can obtain
meaning only by being related to the terms of some other point of
view that is familiar and known by all' ([1963d], p. 296). But if that

were the case, how could the latter point of view and *its* language
have become familiar? Unless they did so without being connected to
any previous language, a vicious infinite regress looms. But if everyday language did arise without connection to a previous language,
why should a new language not arise in the same way? Feyerabend
identifies the guiding idea behind the objection as the view that 'new
and abstract languages cannot be introduced in a direct way but must
first be connected with an already existing, and presumably stable,
observational idiom' ([1970a]: *PP2*, p. 155). This idea that the terms of
a theory receive their interpretation indirectly, by being related to a
different conceptual system which is an older theory or observation-
language, is, as Feyerabend says, a fundamental principle of logical
empiricism. For Carnap, Feigl and Nagel, Feyerabend claims, older
theories or observation-languages are adopted not because of their
theoretical excellence (in fact, they have been refuted) but simply
because they are used by a certain language community as a means
of communication. But this idea is refuted at once, he suggests, by the
way in which children learn to speak their first language, and by the
way in which anthropologists and linguists learn the unknown
languages of newly discovered tribes. What is overlooked by the
logical empiricist objection, says Feyerabend, 'is that an initially
ununderstood alternative may be *learned* in the way in which one
learns a new and unfamiliar language, not by *translation*, but by *living*
with the members of the community where the language is spoken'
(*PP2*, p. 155 n. 58). This idea is, he claims, for very good reasons,
anathema in linguistic and anthropological field-work.

This particular attack on logical empiricism is unpersuasive. The
defensible core of the objection Feyerabend is considering is that no
theoretical language can be understood without its being connected
with an observation-language. This idea is *not* refuted by the way(s)
in which linguists, anthropologists and children learn new languages, for the languages they learn, 'everyday' languages, are more
akin to observation-languages than to the purely theoretical languages Feyerabend has in mind (even though they will contain terms
for unobservables, such as the term 'mind').

We have already seen, in chapter 2, that theoretical statements do
still retain a crucial semantic connection with observation. If theoretical statements are meaningful, it must be possible to learn and to
explain their meaning. In order to do this, there must be situations in
which these statements are used correctly. But in order to explain to
someone learning the language the situations in which its statements
are used correctly, these situations must be recognizable. They must,

therefore, be identifiable within our ordinary material-object language. The 'direct' introduction of theoretical terms which Feyerabend has in mind is impossible.

In fact, the way Feyerabend relates this argument betrays the fact that there is another ready reply to it. It is observation-languages with which scientific theories are semantically connected, and *languages* cannot be refuted. At most they may become useless for certain purposes. However, as Strawson suggested, where the observation-language is the material-object language itself, it cannot even suffer this fate. Our material-object language is here to stay. It is therefore not unreasonable to demand that new theoretical terms and statements bear some (perhaps indirect) connection with terms and statements of our material-object language. The 'choice' of ordinary material-object language as our observation-language or of an older theory as a basis for interpretation is not (*pace* Feyerabend) due to its popularity, but rather to its necessary role at the heart of our conceptual scheme.

8.3   Reductive Materialism

As early as 1958, Feyerabend thought his version of scientific realism could resolve the mind–body problem. His first idea was that that problem arises when we interpret a single set of physiological processes in terms of two observation-languages which contain very different theories. The result will be two sets of apparently different phenomena. Pains and bodily injuries, for example, are phenomena which may result from interpreting a single set of physiological processes in terms of two theories, one phenomenological, and one materialistic. Feyerabend's scientific realism then advises that a materialist interpretation of the languages in which we talk about 'pains' and 'injuries' may show that these terms denote one and the same set of physiological processes: two very different observation-languages, he says, can be 'united and jointly interpreted by one and the same theory' ([1958a]: *PP1*, p. 32).

Feyerabend soon extended this line of thought, claiming explicitly that mental terms and terms for associated physiological processes are incommensurable. If this is so, he said,

> it is of course completely impossible either to reduce them to each other, or to relate them to each other with the help of an empirical hypothesis, or to find entities which belong to the extension of both kinds of terms. That is, the conditions under which the mind–body problem has been set up as

well as the particular character of its key terms are such that a solution is forever impossible: a solution of the problem would require combining what is incommensurable without allowing for a modification of meanings which would eliminate this incommensurability. ([1962a]: *PP*1, p. 90)

This account of the matter is clearly untenable. For mental terms to be truly incommensurable with physiological terms, psychological statements and physiological statements would have to be (or embody) rival accounts of the *same* phenomena. But those who argue that psychological statements are irreducible to statements about material objects presuppose that these are not just rival accounts of the same things. The suggestion that these two kinds of terms are incommensurable is of no help with the mind–body problem, since in order for these terms to be incommensurable that problem would already have to have been settled in favour of the materialist!

As is so often the case, Feyerabend then tried to turn the argument into a dispute over theoretical freedom:

All these difficulties disappear if we are prepared to admit that, in the course of the progress of knowledge, we may have to abandon a certain point of view and the meanings connected with it – for example if we are prepared to admit that the mental connotation of mental terms may be spurious and in need of replacement by a physical connotation according to which mental events, such as pains, states of awareness, and thoughts, are complex physical states of either the brain or the central nervous system, or perhaps the whole organism. (*PP*1, p. 90)[2]

He then used the concept of incommensurability to give a general diagnosis of the nature of philosophical problems, a diagnosis which contains the nerve of his critique of 'linguistic philosophy'. Linguistic arguments, he claimed, are obstacles to theoretical progress, since they amount to the pernicious demand for meaning invariance. When the scientific philosopher seeks to identify A's with B's, his linguistic opponent may remind him that not only is 'A' not synonymous with 'B', but A's have properties which B's do not have, or vice versa. In the course of arguing that any form of meaning invariance leads to difficulties in giving an account of the growth of knowledge or establishing correlations between entities described with the help of incommensurable concepts, Feyerabend remarked that:

these are exactly the difficulties we encounter in trying to solve such age-old problems as the mind–body problem, the problem of the reality of the external world, and the problem of other minds. That is, it will usually turn out that a solution of these problems is deemed satisfactory only if it leaves unchanged the meanings of certain key terms and that it is exactly this

condition, the condition of meaning invariance, which makes them insolu-
ble. . . . [T]he demand for meaning invariance is incompatible with empiricism. Taking all this into account, we may hope that once contemporary
empiricism has been freed from the elements it shares with its more
dogmatic opponents, it will be able to make swift progress in the solution
of the above problems. (*PP*1, p. 47)

Although this line of thought contains the important recognition that
philosophical problems are conceptual, it is hard not to see the rest of
it as confused. To try to deal with philosophical problems by initiating
a conceptual change will just ensure that those problems are left
behind unsolved. Conceptual problems cannot be *solved*, but only
evaded, by conceptual change.

Having said this, we must recognize that the solution (or dissolu-
tion) of the mind–body problem is not Feyerabend's main concern.
Rather, the impetus behind his view is the desire to make progress, to
improve our theoretical account of what some materialists now refer
to as 'the mind/brain'. Feyerabend is explicit about this. At the very
beginning of his major paper on the mind–body problem, he announces his intention of putting philosophy 'in its proper place':

> It occurs only too often that attempts to arrive at a coherent picture of the
> world are held up by philosophical bickering and are perhaps even given
> up before they can show their merits. It seems to me that those who
> originate such attempts ought to be a little less afraid of difficulties; that
> they ought to look through the arguments which are presented against
> them; and that they ought to recognise their irrelevance. . . . To encourage
> . . . development from the abstract to the concrete, to contribute to the
> invention of further ideas, this is the proper task of a philosophy which
> aspires to be more than a hindrance to progress. ([1963b]: *PP*1, p. 161)

And at the very end of the same paper, he sets out the proper task of
philosophy thus:

> [T]he task of philosophy, or of any enterprise interested in the advance
> rather than the embalming of knowledge, is to encourage the development
> of . . . new modes of approach, to participate in their improvement rather
> than to waste time in showing, what is obvious anyway, that they are
> different from the status quo. (p. 175)

What ultimately underlies Feyerabend's materialism, therefore, is his
strongly scientistic conception of philosophy.

It is true that in the course of scientific change it is necessary
to abandon both certain theories or points of view and certain terms,
languages, concepts, or means of representation. So no 'condition of

meaning invariance' can be enforced in science: scientists are free to indulge in both theoretical *and* conceptual change. But the nature of the change is not the same in each case. Theories are truth-claims. When we abandon a theory in favour of a better theory, therefore, it makes sense to say that we have made progress: in discovering that what we previously believed was false, we move closer to the truth. But languages and their constituent concepts are not, and do not involve, truth-claims. Rather than being attempts to describe reality (or anything else), languages and concepts are preconditions for any such description. To assimilate languages and concepts to theories, and to assimilate conceptual change to theoretical change, as Feyerabend constantly does, is therefore superficial. When we abandon a language or a concept, we cannot be said to have progressed in the same way as when we abandon a theory. By 'progress' Feyerabend usually (during this phase of his work) means progress towards the truth. But when he appeals to 'progress' in connection with conceptual, rather than theoretical change, it is not clear what he can mean other than change for its own sake.

One of Feyerabend's most basic errors is to think that one can examine and evaluate problems in one conceptual framework by changing to a very different framework. Just as changing concepts does not solve problems formulated in terms of those concepts, so problems in one framework cannot be *solved*, but only swept under the carpet, by switching to another. The 'condition of meaning invariance', as applied to conceptual problems, is just a reminder that we cannot solve such problems by changing the subject. It therefore has a secure place within philosophy for as long as philosophers do anything like 'conceptual analysis'. In sum, Feyerabend's attack on 'linguistic philosophy' is, I think, weak. He did not successfully rebut the linguistic philosophers' objections to materialism.

8.4 Eliminative Materialism

In the first of his two 1963 papers on the mind–body problem, Feyerabend changed tack. He considered the then-fashionable materialist 'identity hypothesis', which sought to identify mental phenomena with physical phenomena. Such identity hypotheses are the favoured instruments of reductive versions of materialism: views which attempt to reduce the mental to the physical. Different identity hypotheses can be formulated in terms of substances, processes, events, states or properties. Feyerabend focused on an identity hypothesis for processes:

(H): X is a mental process of kind A if and only if X is a central nervous system process of kind α.

Instead of defending such hypotheses by appealing to scientific progress, he now urged that materialists should move beyond them. Identity hypotheses, Feyerabend said, backfire, since although they are meant to imply that mental processes are really physical, they can equally be read (from right to left) as asserting that (some) physical processes (viz., processes in the brain or central nervous system) have non-physical properties. Instead of replacing dualism with materialism, as they intended, materialists would thus replace one form of dualism (process-dualism) with another: property-dualism. In defending identity hypotheses, Feyerabend suggested, the materialist's commitment to monism, the idea that all substances, processes, events, states and properties are physical, has been sold out. If monism is true, Feyerabend held, identity hypotheses must be false. In this case, instead of there being mental processes which are identified with physical processes, 'there are then *no* mental processes in the usual (non-materialistic) sense' ([1963d], p. 295). Feyerabend therefore advised materialists to develop their theory without recourse to existing mental terms. This is the first flowering of an alternative to reductive materialism, now known as *eliminative* materialism. This very radical and counter-intuitive version of materialism can be seen as an application of super-realism.

In fact, Feyerabend's argument should not persuade. An identity hypothesis such as H does not imply that some physical processes have non-physical features. On its own, it says nothing about mental or physical features (properties) at all. While such an identity hypothesis is compatible with property-dualism, therefore, it does not imply that view. Feyerabend's deeper worry seems to be that such a hypothesis implies the existence of mental processes. But that should not worry the materialist, even the materialist monist since, although the hypothesis does indeed assert the existence of mental processes, it identifies them with (a subclass of) things the materialist already believes in, namely, physical processes. The fact that such an identity hypothesis uses the term 'mental process' in its 'usual (non-materialistic) sense' does not (yet) mean that it implies the existence of anything non-physical at all.

Property-dualism involves the untenable idea that psychological properties are properties of the brain (when in fact they are properly predicable only of the *person*). But even if property-dualism is unacceptable, the materialist monism which could be embodied in a complete range of identity hypotheses has also proven hard to defend. Insofar as we have a clear conception of which properties are physical properties, it is implausible to hold that those properties we think of as psychological (those picked out by what Strawson called 'P-predicates') are physical properties. Aside from a few clear examples of physical properties (mass, length, shape, etc.), we simply have no conception of which other properties *are* physical properties. Perhaps this is why the eliminativist strategy has tempted materialists: it promises to do away with all philosophical disputes about categories, by placing everything in the category of the physical and simply denying the existence of any thing (property, state, event, etc.) that does not fit into this category.

Feyerabend's philosophy of mind came to fruition in his second 1963 paper, 'Materialism and the Mind–Body Problem', where he set out to defend materialism against a certain kind of attack. Although this paper is now thought of as a *locus classicus* of eliminative materialism, it is well to note that its eliminativism coexists (uneasily) alongside a defence of the identity hypothesis which embodies the reductive materialist view. I shall give a critical running commentary on the highlights of this paper.

Materialism is conceived of here simply as the view that 'the only entities existing in the world are atoms, [and] aggregates of atoms and that the only properties and relations are the properties of, and the relations between, such aggregates' ([1963b]: *PP1*, p. 161). The envisaged attack on it comes in two parts. The first objection is that materialism cannot give a correct account of human beings because, apart from being material, humans also undergo mental processes (experiences, thoughts, pains, etc.) which cannot be analysed in a materialistic way. The claim that experiences, thoughts, pains, etc., are not material processes is then supported by pointing out that '[t]here are statements which can be made about pains, thoughts, etc. which cannot be made about material processes; and there are other statements which can be made about material processes but which cannot be made about pains, thoughts, etc.' (p. 162). This latter impossibility is supposed to obtain either because such statements would be meaningless, or because they would be false. Feyerabend therefore divides this first objection into two arguments.

The first argument, from meaninglessness, is the one preferred by linguistic philosophers. They stress that although brain processes have certain features, like physical location, it makes no sense to speak of certain mental phenomena as being physically located. Since this argument is based on 'grammar' (roughly, semantical rules determining what it makes sense to say), Feyerabend opens his riposte by deploying his contextual theory of meaning:

> Whether or not a statement is meaningful depends on the grammatical rules guiding the corresponding sentence. The argument appeals to such rules. It points out that the materialist, in stating his thesis, is violating them. . . . It is evident that this argument is incomplete. An incompatibility between the materialistic language and the rules implicit in some other idiom will criticise [sic] the former only if the latter can be shown to possess certain advantages. (p. 162)

So Feyerabend concedes the original point that the materialist thesis violates the grammatical rules of our ordinary language, but he thinks it does not matter, since what is needed is a debate about which is the *better* language. Some languages, he thinks, can be shown to be scientifically inadequate.

The objector should, I think, deny that this is the issue. When the materialist seeks to identify something picked out by terms of ordinary language with something physical, the issue is not which of the two languages is better: unless the original ordinary language identification of mental phenomena is correct to begin with, materialists do not even have the resources to state the identity hypothesis in which they try to embody their view. In this way, materialists who want to give an account of mental phenomena at all *presuppose* the integrity and meaningfulness of our ordinary psychological language, whether they like it or not. This is perhaps another of the considerations that has tempted materialists away from identity hypotheses and towards the eliminativist version of materialism. That version promises to liberate them from what they see as an outdated and impoverished way of identifying the real, underlying neural data that we should be explaining.

Nevertheless Feyerabend presses on, to discuss the comparative merits of 'ordinary language' and a materialist language. The fact that the former is in common use he deems an irrelevant historical accident. The only other merit of ordinary language that has been suggested is its practical success. Feyerabend takes aim at the following passage from J. L. Austin:

> Our common stock of words embodies all the distinctions men have found worth drawing, and the connections they have found worth marking, in the lifetime of many generations: these are surely likely to be more

numerous, more sound, since they have stood up to the long test of the
survival of the fittest, and more subtle . . . than any that you or I are likely
to think up. (Austin [1979], p. 182)

This argument Feyerabend compares to the orthodox view, within
the philosophy of science, that highly confirmed theories are preferable to less highly confirmed ones. Austin's argument is less plausible
than this, Feyerabend thinks, since scientific theories are testable,
whereas ordinary language is not. But even if this were not so, and the
parallel was legitimate, Feyerabend argues that a preference for
highly confirmed theories is misguided. Against such a preference,
he wields his argument for theoretical pluralism: we can find the
weaknesses of an all-pervasive theory such as ordinary language not
by comparing it with observations or facts, but only by comparing it
with *other* languages. Therefore we should invent and develop other
languages, including the envisaged materialist language. Only when
this materialist language is fully developed can we legitimately
compare its virtues with those of ordinary language. In the meantime,
the materialist must be allowed to develop his or her language, and
the point of view it allegedly contains, free from hindrance by the
linguistic philosopher.

Analytical philosophers should respond to this line of thought by
reiterating that philosophical problems cannot be solved by conceptual change. The original philosophical problem concerns the relationship between mental and physical concepts. *That* problem cannot
be solved by comparing the merits of our psychological language
with those of a materialist language, and then rejecting the former. Of
course, there would be no philosophical mind–body problem left if
we *were* to reject all our psychological language. But this would be
because we had ignored the problem, not because we had solved it.
None of this means that the development of a materialist language is
in any way forbidden. Such a language might come to be widely used
(although it seems inconceivable that it should displace our ordinary
language altogether, that our descendants should simply cease to use
any psychological concepts). But none of this would establish the
sought-for conclusions that such a language is better than ordinary
language (better at what?), or that the widespread use of a materialist
language would mean that the mental had been shown to be physical.
For that to be shown, the two languages must be usable in parallel to
refer to the very same things. Not only has Feyerabend not shown that
this is possible, but he has not shown how to rebut the objection that
since there are things it makes sense to say about psychological
phenomena which it does not make sense to say about physical

phenomena, and vice versa, the former cannot be literally identical with the latter.

So much for the argument from meaninglessness. The second argument Feyerabend considers says that the materialist's identity hypotheses are meaningful, but false. Here, it is supposed that we can establish by observation that thoughts (experiences, pains, etc.) cannot be brain processes, or material processes of any kind. The idea is that, whereas brain processes have certain properties, introspection reveals that thoughts do not have such properties, or vice versa.[3] In response to this, Feyerabend appeals to his scientific realism, and seeks to deploy the distinction between appearance and reality. Introspection, he thinks, gives us access only to the appearances of mental processes. So the fact that introspection reveals differences between brain processes and the appearances of mental processes fails to show that mental and physical processes really differ. We can still postulate an identity between what is opaquely revealed in introspection and a neural process. A language based on such a postulate, says Feyerabend, 'would differ significantly from ordinary English. But this fact can be used as an argument against the identification only *after* it has been shown that the new language is inferior to ordinary English' ([1936b]: *PP*1, p. 166). Impatient with those who would reject it, Feyerabend then argued that realism should not be confined to processes outside one's skin. This is true: if we are going to be scientific realists about physics, we ought to take the same attitude towards postulated neural processes as well. But this will not work as an argument for realism about the mental: it does not establish that we ought to be realists about mental phenomena unless we have already granted that mental phenomena *are* processes inside one's skin. But this is just the point at issue. One who denies that mental phenomena are brain processes should also deny that they are bodily processes of any kind. (If the materialist's opponent were to concede that mental phenomena are processes inside the skin, what could possibly prevent him or her from identifying them with brain processes, or, more widely, processes in the central nervous system? Surely not the idea that they are *other* physiological processes?) To say that mental processes are 'inside one's skin', if it is not to endorse materialism, is a mere metaphor.

Feyerabend then considers a response on the part of the materialist's opponent: it might be said that in the case of thoughts, sensations, and feelings the distinction between appearance and reality does not apply. The idea behind this response Feyerabend represents as follows:

Mental processes are things with which we are directly acquainted. Unlike physical objects, whose structure must be unveiled by experimental research and about whose nature we can make only more or less plausible conjectures, they can be known completely, and with certainty. Essence and appearance coincide here. (*PP1*, p. 167)

Feyerabend, unlike most contemporary materialists, seems disinclined to dispute this. He does not seek to defend the idea that one can be wrong about one's own mental phenomena. Instead, he accepts that we do have incorrigible knowledge of at least some of our mental lives, but goes on to ask *why* we should be thus incorrigible. His answer is that statements about mental processes are logically incorrigible because they lack empirical content. In this respect, they compare unfavourably with material-object statements:

Statements about physical objects possess a very rich content. They are vulnerable because of the existence of this content. Thus, the statement 'There is a table in front of me' leads to predictions concerning my tactual sensations; the behaviour of other material objects; the behaviour of other people, etc. Failure of any one of these predictions may force me to withdraw the statement. This is not the case with statements concerning thoughts, sensations, feelings; or at least there is the impression that the same kind of vulnerability does not obtain here. The reason is that their content is so much poorer. No prediction, no retrodiction can be inferred from them, and the need to withdraw them can therefore not arise. (pp. 167–8)

Feyerabend then thinks it appropriate to investigate the source of the rich content possessed by material-object statements. In line with his normative epistemology, he argues that material-object statements are richly falsifiable, and psychological statements unfalsifiable, only because we have decided to make them so. In other words, he turns our attention from the descriptive issue (are psychological statements incorrigible?), on which he concedes defeat, towards a normative issue (are we *right* to have made psychological statements incorrigible?). It is on this normative issue that he proposes to make a stand, suggesting that we should enrich the content of psychological statements, infusing them with our considerable knowledge about the mental, in order to *make* them falsifiable. This, he urges, would be in line with the demands of scientific methodology.

This line of argument, as well as the objection to materialism to which it responds, embodies errors about the nature of our first-person authority. Its not making sense to doubt certain of one's own sincere first-person psychological characterizations (like 'I feel cold', or 'I think I see an elephant') should not be conceived as showing that one has incorrigible knowledge of certain matters of subjective fact. Such ascriptions lack empirical content not because they are incorrigibly known, but rather because they are avowals, which cannot be said to be known at all.[4] Feyerabend's suggestion that we could replace such avowals with (fully corrigible) hypotheses fails to recognize that the *use* of avowals, their function, differs categorically from that of hypotheses. Their lack of empirical content is not a feature which could possibly be remedied by turning them into hypotheses, consistently with them retaining their current, expressive function.

The third and final anti-materialist argument Feyerabend considers is the traditional empiricist refrain that we can know our own pains, thoughts, feelings, etc., *by acquaintance*. The idea is that in having these items present to one's mind, one knows them both directly (without inference) and completely (through and through), thereby discovering that they are not material processes. This argument Feyerabend deems circular:

> If we possess knowledge by acquaintance with respect to mental states of affairs, if there seems to be something 'immediately given', then this is the *result* of the low content of the statements used for expressing this knowledge. Had we enriched the notions employed in these statements in a materialistic (or an objective-idealistic) fashion *as we might well have done*, then we would no longer be able to say that we know mental processes by acquaintance. (p. 170)

The alleged 'fact' of knowledge by acquaintance may indeed be a fact. But even so, Feyerabend suggests, it would still be only 'the result of certain peculiarities of the language spoken *and therefore alterable*' (p. 171). Materialism recognizes that this 'fact' is alterable, and suggests that it be altered. So materialism cannot be refuted by reference to this fact: what must be shown is that its *suggestion* is undesirable, and that acquaintance is desirable.

We should not get the impression that Feyerabend is granting materialism any special licence in this respect: he would allow that any general theory has the right to ignore (or even to 'alter') the 'facts' which are posited and explained by other theories. His argument aims only at granting materialism an indefinitely long breathing-space during which no empirical objections are granted decisive weight. The same would apply to materialism's rivals. During this phase of the debate, the only way of disposing of a theory is to show that it conflicts with our finest *methodological rules*. Insofar as a theory posits the existence of incorrigible knowledge, whether of observation-statements or of *a priori* principles, it will be regarded as meth-

odologically pernicious, and therefore dispensable. Feyerabend thinks
all versions of dualism will fall at this hurdle. They will not make it
to the final hurdle, at which the substantive scientific virtues of the
surviving methodologically acceptable theories are compared.

Feyerabend sees in the argument from acquaintance an example of
a baneful philosophical strategy: to present as a fact about the mind
something which is merely 'the result of the way in which any kind
of knowledge (or opinion) concerning the mind has been incorporated, or is being incorporated into the language used for describing
facts' (p. 171). This he considers to be another example of
instrumentalism. Instead of using our knowledge about the mind to
enrich our psychological concepts, as the realist proposes, we use it to
make predictions in terms of the existing, impoverished concepts.
The alleged fact, says Feyerabend, is therefore merely 'a projection,
into the world, of certain peculiarities of our way of building up
knowledge' (p. 171). It presents the result of our *decision* to adopt an
instrumentalist interpretation of psychological language as a fact of
nature. Philosophers proceed in this way, he suggests, because all of
them, apart from Wittgenstein,[5] hold a philosophical theory according to which:

> the world consists of two domains, the domain of the outer, physical world,
> and the domain of the inner, or mental world. The outer world can be
> experienced, but only indirectly. Our knowledge of the outer world will
> therefore forever remain hypothetical. The inner world, the mental world,
> on the other hand, can be directly experienced. The knowledge gained in
> this fashion is complete, and absolutely certain. (p. 171).

While this is undoubtedly a familiar philosophical picture of the
relationship between the self and 'the external world', one which
figures prominently in the empiricist tradition from which the argument from acquaintance springs, Feyerabend correctly considers it
wildly implausible to think that it resides within our psychological
language. He complains that instead of defending it explicitly, thus
making it available for criticism, philosophers, in reading this theory
into our language, insinuate that it is something we all share and
simultaneously disguise the fact that *they* endorse it. This is tantamount to an instrumentalist stratagem: insulating the theory from
possible falsification by failing to present it as a *theory* at all.

Feyerabend is here relying on his view that factual or empirical
arguments are worth little. He can mobilize several considerations in
this regard. First, there is the possibility that a theory's empirical
success has been manufactured as the result of a myth predicament.[6]

Connected with this is the second idea that apparently empirical arguments may well be circular: they may create 'facts' which appear to support a particular theory.[7] Third, empirically adequate observation-statements may have to be reinterpreted, not because they do not adequately express what is perceived, but because of changes elsewhere in the conceptual system to which they belong.[8] Our conception of the mind may have impressive confirmation, but such evidence is weak: 'we should not to be too impressed by empirical arguments but . . . should first investigate the source of their apparent success' ([1963b]: *PP1*, p. 172).

This argument against classical empiricism has, I think, too many loopholes. Classical empiricists could justifiably respond that it has not been shown *how* knowledge by acquaintance is the result of our investing psychological statements with little empirical content, and that the possibility of 'enriching' these statements does *not* mean that knowledge by acquaintance does not exist. In the very final analysis, when it comes to arguing on the terrain of methodology, classical empiricists can simply dig their heels in and insist that Feyerabend's proposed rules are *not* rules of acceptable scientific methodology, since they fail to underwrite the alternative ideal of sense-certainty to which classical empiricists subscribe.

Feyerabend's defence of materialism must, I believe, be judged ineffective. He does not succeed in solving or removing the conceptual problems inherent in the relationship of mind to matter. He fails to show that psychological concepts are incommensurable with material-object concepts, since he fails to show that they conceptualize the *same* phenomena. His diagnosis of the nature of philosophical problems and his conception of the task of philosophy are problematic, his arguments against linguistic philosophy ineffectual. Although he started out wanting to defend an 'identity hypothesis', characteristic of reductive materialism, he misunderstood the nature of such hypotheses, leaping prematurely into an eliminative materialist view. The materialist monism he endorsed he also left undefended. Along with other materialists, Feyerabend failed to rebut the objection that identity hypotheses are attempts to identify items which are in different logical categories. Transmuting the argument into a dispute about which language is methodologically most acceptable evades the issue, and makes no sense. Unless our ordinary psychological language is in order, reductive materialism cannot even be stated. To switch to the uncompromising eliminativist line here is to ignore the conceptual problem, not to solve it. Finally, even Feyerabend's attack on the classical empiricist appeal to knowledge by acquaintance was less effective than his usual indictments of empiricism. Our concluding verdict must be that he did not show that identity hypotheses are defensible, or that eliminative materialism is either a reasonable way of defending them or a reasonable alternative to them. His arguments should not convince us that the best route from reductive materialism is onward into a more radical, eliminative view.

8.5 'Folk Psychology'

Feyerabend's defence of materialism, whether reductionist or eliminativist, was unconvincing. However, the approach to philosophy of mind which he staked out has come in for much attention in recent years, and we ought to consider whether his view can be reconditioned to provide a better argument for eliminative materialism.

Feyerabend's materialism focused on mental *events* (his examples being pains, sudden thoughts, feelings, states of awareness, and momentary experiences). His successors within the philosophy of mind, Paul and Patricia Churchland, seem to be reductionist about most of these.[9] But the Churchlands have refreshed the argument for eliminative materialism, directing it instead at those psychological phenomena picked out by what Bertrand Russell christened 'verbs of propositional attitude'. Verbs such as 'to believe', 'to intend', 'to desire' and 'to hope' are used to ascribe attitudes which exhibit the feature known as intentionality: they are about, or directed at objects or states of affairs, but they do not depend for their existence on the existence or occurrence of the objects or states of affairs specified as the content of the attitude. So, for example, one can believe that p without it ever being the case that p, one can intend to φ without ever φ-ing, one can fear the creature from the Black Lagoon without there being such a creature. Taken together, verbs like these are supposed to specify a domain of mental phenomena now known as 'Folk Psychology' (henceforth, often, 'FP').

Paul Churchland developed the concept of Folk Psychology from Sellars' 1956 essay 'Empiricism and the Philosophy of Mind', in which it was suggested that the concept of mind was 'the germ of a theory' (Sellars [1963], p. 187), and something which would, in evolutionary history, have improved its users' ability to explain and predict the behaviour of others, thus conferring upon those users a distinct survival advantage. But Sellars wavered on the issue of whether psychological concepts are truly *theoretical* concepts.

Churchland goes beyond Sellars in claiming explicitly that our

common-sense conceptual framework is theoretical, and that the part
of it which concerns the concept of a person (Folk Psychology)
constitutes a *theory* of (part of) our psychological functioning. To
characterize FP as a theory is, among other things, to say that its
integrity depends on its ability successfully to explain and predict the
behaviour of human beings (oneself included). To think of other
people as having minds is, on this view, an inference to the best
explanation: the hypothesis that others have mental states is the best
explanation of why they behave as they do. Churchland recognizes
that FP must therefore be shown to have the features normally
regarded as characteristic of theories. He attempts to meet this
obligation by outlining generalizations familiar from our explanatory practice, the principles of a common-sense theory of the determi-
nants of human behaviour and which he identifies as the putative
laws of Folk Psychology. These law-like statements are to be regarded
as confirmed insofar as they allow us to predict and explain behaviour. To account for our ability to explain, predict and understand
human behaviour in terms of mental states, says Churchland, we
must suppose that we share 'a command or tacit understanding of a
framework of abstract laws or principles concerning the dynamic
relations holding between causal circumstances, psychological states,
and overt behaviour' (P. M. Churchland [1979], p. 92). This proposal,
he claims, supplies us with a unified account of many of the problems
facing the philosophy of mind (such as the problem of other minds,
the nature of introspection, intentionality, and the open-endedness of
our criteria for specific mental phenomena). But if we embrace this
proposal the mind–body problem is itself transformed: it becomes the
problem of the relationship between the ontology of one theory (FP)
and that of another (our general theory of human neurophysiology).
Then, in light of the fact that the latter theory may threaten the former
with outright replacement, we are forced to take seriously the possibility that all our 'introspective judgements' may be systematically
false, that there may be no folk-psychological phenomena, and no
persons (as that term is usually understood). This is full-blooded
eliminative materialism.

Although Feyerabend did not use the concept of Folk Psychology,
it fits comfortably into his way of thinking. His early remarks on the
mind–body problem, after all, assimilate our conception of the mental to a language which contains a certain theory. But there is nevertheless one central respect in which he would not have accepted the
account of FP common to many contemporary mentalists and materialists. These philosophers portray Folk Psychology as burdened

with dualism. They think of Folk Psychology as an empirical theory whose explanatory ambitions are furthered by postulating the existence of 'inner' mental states, events and processes which stand in causal relations to one another and to behaviour. Mentalists (like Jerry Fodor) then argue that these mental phenomena really exist, and that each of their instances can be identified with neurophysiological phenomena; whereas eliminativists argue that the phenomena postulated by FP are so different from neural goings-on that they must be supposed not to exist at all. Feyerabend, as we have seen, recognized this as a misinterpretation of our psychological language to which most, but not all, philosophers have succumbed. His realization that philosophical dualism does not lie within our psychological language, but is imposed upon it by philosophers, makes his view superior to those of his materialist successors.

Using the concept of Folk Psychology, we can now outline the basic argument for eliminative materialism:

1 Scientific theories are our best guesses about the nature of any domain of empirical phenomena. We have an epistemic duty to conceptualize any such domain in what we believe to be the best way we can, and this means deploying our best scientific theories.

2 There are scientific theories of the mind (henceforth, STMs), which can be roughly compared with one another in terms of their overall scientific virtues (predictive and explanatory power, scope, coherence, simplicity, etc.).

3 There is a common-sense theory of the mind, Folk Psychology, which all of us use and most of us (perhaps all of us) endorse, but which is not necessarily true.

4 FP and our best STM are rivals. They cannot both be true.

5 FP and our best STM may not be semantically commensurable, but their overall scientific virtues can be roughly compared.

6 Such comparison would reveal that FP is notably less impressive than our best STM.

7 There is *something* we (some of us, at least) can and should do about our current situation: we can and should try to cease thinking of ourselves in terms of FP, and to reconceive ourselves in terms of our best STM.

Almost every step in this argument can be challenged. Many, under the influence of worries about incommensurability, will resist the suggestions (2 and 5) that we can (even roughly) compare the virtues of very different scientific theories or, *a fortiori*, that we can compare any scientific theory with any common-sense 'theory' such as Folk Psychology. This complaint might seem particularly pressing against claim 5, where a diagnosis of incommensurability might look promising. Eliminativists may respond by claiming, as Feyerabend would once have done, that even incommensurability would not preclude comparison. All it precludes, on this view, is what Paul Churchland calls the 'point-by-point semantic comparison' of theories. Such comparison, it would then be argued, is not necessary for an overall estimate of scientific merit. The trouble with this line of argument, I believe, is that the relationship between scientific theories and Folk Psychology is not even close enough to be characterized as incommensurability. This becomes clear in considering claim 4.

Claim 4, that FP and our best STM are in competition, certainly seems vulnerable. In the mind–body problem we have no conception of any *single domain* which is being differently described by both 'Folk Psychology' and Churchland's favourite STM, which is neuroscience. Even if we grant that FP is a theory, and a bad theory, one would be hard-pressed to argue that it is a bad theory *of the brain* and its functioning. (Aristotle, after all, could employ 'Folk Psychology' while thinking that the brain was simply an organ for cooling the blood.) But having a common subject matter is a precondition for rivalry of theories, and therefore for their incommensurability. (It is significant, in this connection, that Patricia Churchland, who needs FP to be a rival to neuroscience, *invents* a common domain: the 'mind/brain'! This simply *presupposes* an identity hypothesis which is incompatible with non-materialist views *and* with her own eliminativism.) This objection will, of course, leave untouched those who think of FP as a rival to some higher-level theory, such as cognitive psychology, which can straightforwardly be said to be about the *mind*. All it means is that while there may be scientific theories of the nature of the mind, neuroscience is not one of them.

Recall that one general problem with the concept of incommensurability is that the more radically theories differ, the less plausible it becomes to think of them as rival accounts of the same phenomena. If we accept that they are rivals, they must have enough in common to license this judgement. If this condition is supposed to be satisfied in the mind–body case, we are already committed to an identity hypothesis, and thereby to reductive materialism. But in fact the assumption that FP and STMs (where this includes neuroscience) are rival accounts of the same phenomena was never substantiated in the first place. (That, it might be said, is one reason why the mind–body problem is a *conceptual* problem).

Claim 3 has been attacked on the grounds that Folk Psychology is not a *theory*.[10] This idea, although widely associated with eliminativism, may not seem essential to it. An eliminativist, it may seem, could concede (a) that some statements of FP are 'analytic' or conceptual truths which collectively constitute our common-sense *concept* of the mind. Concepts are not evaluable as true or false, and are not directly at the mercy of incoming empirical information. The eliminativist still has room to wield the crucial contention that we can isolate some other subset of FP statements that *are* empirical and thus corrigible, claims whose falsity would be enough to condemn FP to the dustbin of history. The further concession (b), that such psychological claims are not *theoretical*, also seems reasonable in the light of objections to the misuse and trivialization of the term 'theory'. What remains essential to eliminativism, on this response, is that there should be some body of common-sense claims about the mind and its operation which are empirical and genuinely corrigible. While common-sense psychological *concepts* do not constitute a theory, common-sense psychological *beliefs are* empirical (although not theoretical), and are therefore fallible.

The trouble with this response, from the eliminativist's point of view, is that it threatens to skew the debate about the integrity of FP towards discussion of singular psychological claims and the negative existential thesis of eliminativism, that there are no such things as beliefs, desires, intentions, etc. Avowals like 'I believe that Paris is the capital of France', 'I hope to survive beyond the year 2000', and 'I want an ice-cream' cannot be represented as corrigible *or* incorrigible. Their obvious propriety casts serious doubt upon the claim that they owe their integrity to a theoretical context. But if they are in order, there cannot be anything wrong with the concepts in terms of which they are framed, the concepts of FP. This platitudinous nature of FP's existential claim, that there are such things as beliefs, desires, etc., is, I believe, the reason why eliminativists have clung to the supposition that the 'laws' of FP are theoretical.

Others have conceded that FP is a theory, a theory which is in competition with and can be compared with STMs, but have protested against the supposition (6) that FP would suffer in the comparison. Insofar as this response has force, I think it derives largely from pointing out the truth of the singular psychological claims mentioned above. The idea is that FP is so massively useful, its predictive power so weighty, that it would not come off worse in a fair overall comparison with the merits of any scientific theory of the mind.

Claim 4, which I have already objected to, is not out of the woods even if the eliminativist has a reply to that objection. The claim is that Folk Psychology is incompatible with our best STMs. This is usually argued for by eliminativists as follows: if FP and STMs were to be compatible, then they would either have to be theories of different domains, or theories of the same domain which stand in the relationship of theoretical reduction. In other words, FP would have to be reducible to our favoured STM. But, especially where that STM is neuroscientific, the two theories do not stand in that relation. If FP is irreducible to neuroscience, and yet the two are in competition, the theoretical town ain't big enough for the both of them. And, clearly, if FP suffers in the comparison, it is the one that has to go.

The problem with this argument is that it manifests an impoverished conception of the possible relationships between theories. Even if they are theories of the same phenomena, FP could be compatible with neuroscience without being reducible to it.[11]

Finally, the eliminative materialist conclusion (7) has sometimes been inflated. Rash predictions that the complete overthrow of FP is possible, immanent, or even necessary are sometimes associated with eliminativism. However, neither the demand to abandon mental terms nor the ridiculous prediction that these terms *will* be abandoned are essential parts of the eliminative materialist's case. Feyerabend, it is true, sometimes favoured the hardline idea that mental terms ought to be simply replaced by neuroscience. But there is also a softer line running alongside the uncompromising scientism of the hard line. Here, the idea is that mental terms are to be reconceived, rather than dropped.[12] The real demand is to the effect that if we truly have our own best cognitive interests at heart we should reconceptualize mental phenomena *as* physical phenomena, and that this would inevitably entail helping mental terms evolve towards terms with a purely materialistic content. In short: we can keep the old words, but a meaning-change will be necessary.

Many other objections to the eliminative materialist conclusion have, of course, been proffered. 'Folk Psychology', it is often claimed, is indispensable: we simply cannot conceive of ourselves within the purely descriptive language of physical science, since we would lose our conception of ourselves as agents, speakers, reasoners, etc. The eliminativist conclusion is thus often argued to be pragmatically self-refuting. Just as Strawson's argument showed the irreplaceability of material-object concepts within our conceptual scheme, such objections suggest that something like our basic folk-psychological concepts are irreplaceable too.

9

Science without Method

9.1 'The Stinkbomb'

Feyerabend's best-known work emerged from a long series of letters to his friend Imre Lakatos in which Lakatos defended, and Feyerabend attacked, a constellation of views they referred to as 'rationalism'.[1] Between 1968 and 1971 they formulated several plans for publication of this material, the most long-standing one being that their opposing essays would appear in a single volume entitled *For and Against Method*. Feyerabend published a long paper entitled 'Against Method', overlapping with much of the proposed book and introducing 'epistemological anarchism', in 1970. When Lakatos died unexpectedly in February 1974, Feyerabend was already contracted to produce a text for New Left Books, and he published his material without Lakatos's reply. Feyerabend used to say that *Against Method* (henceforth '*AM*') was a *letter* to Lakatos, not a book, and their correspondence does indeed contain some of what became *AM*. But a more accurate description is the one given in his autobiography: '*AM* is not a book, it is a collage. It contains descriptions, analyses, arguments that I had published, in almost the same words, ten, fifteen, even twenty years earlier' (*KT*, p. 139). In fact, *AM* represents an archaeology of much of Feyerabend's previous work, containing his account of the relation between theory and experience, his objections to theoretical monism and cumulativism, the Brownian motion example, the central argument for theoretical pluralism, the thesis of incommensurability, and many other things already discussed here. In this chapter, therefore, I shall concentrate on the new and substantive material *AM* has to offer.

Long before its publication, Feyerabend anticipated that *AM*, which he began referring to as 'the stinkbomb', would infuriate people. He was well aware of its inadequacies, and of the inflammatory nature of some of the offhand comments it contained. In a letter of March 1970, he confessed to Lakatos that 'there are too many wilful contradictions in the whole stuff' to get him 'a following of serious people'. At times, he did not even want his 'anarchistic' arguments to succeed. Epistemological anarchism would be, he said, 'just another passing stage' in his life. Lakatos seems to have had much the same idea, saying: 'I do not think that we can drag out this game [viz. 'rationalism' vs. 'anarchism'] much longer, and I am very keen to have our fun and start some new tune'.

When the book appeared, however, the reviews went way beyond what normally counts as bad press in academic circles. In one review alone (Agassi [1976]), Feyerabend was accused, not entirely unfairly, of extolling lies, conning his readers, making false promises, cheating in his interpretation of Galileo, playing the clown, idolizing totalitarian China, and producing 'hate blasts', violence and vulgarity.

Although Feyerabend really should not have been surprised at the hostile reception accorded to *AM*, it nevertheless seems to have caused him much anxiety and depression (see *KT*, pp. 144–51). He felt it necessary to reply in print to almost all its major reviews.[2] In his autobiography, he admits to having been confused by the critical reactions to *AM*, and having gone on the offensive: 'I was alone, sick with some unknown affliction; my private life was in a mess, and I was without a defense. I often wished I had never written that fucking book' (*KT*, p. 147).

9.2   Epistemological Anarchism

For Feyerabend, the core of 'rationalism' is the thesis that there is something worth calling *the* scientific method, something unchanging which makes all good science science. Because I want to focus on this and not on 'rationalism' as a whole, I shall call this core thesis 'methodological monism'. The heart of *AM*'s case is the claim that methodological monism is false: Feyerabend had become a methodological pluralist. 'Rationalism' certainly meant far more for him, and his own view, which he called 'epistemological anarchism', involved more than mere opposition to this core thesis. But, in retrospect, the moves that make up the rest of epistemological anarchism, the ones that most inflamed the critics, are the weakest parts of *AM*. (That

Feyerabend himself recognized this is testified to by his excising many of them from subsequent editions of the book.) Insofar as his critique went beyond opposition to methodological monism, and insofar as his positive views went beyond methodological pluralism, they seem both more problematic and less important.

AM tries to make plausible[3] the conclusion that 'the events, procedures and results that constitute the sciences have no common structure; there are no elements that occur in every scientific investigation but are missing elsewhere' (*AM³*, p. 1). Its main negative thesis, then, is a limited one: that there is no such thing as *the* scientific method, no single set of methodological rules which scientists have followed and which yield (good) science. Feyerabend hopes to show that

> there is not a single rule, however plausible, and however firmly grounded in epistemology, that is not violated at some time or other ... [S]uch violations are not accidental events, they are not results of insufficient knowledge or of inattention which might have been avoided. On the contrary, we see that they are necessary for progress. (*AM¹*, p. 23, *AM³*, p. 14)

Given any rule, he says, 'however "fundamental" or "necessary" for science, there are always circumstances when it is advisable not only to ignore the rule, but to adopt its opposite' (*AM¹*, p. 23, *AM³*, p. 14). Because it follows no set of rules, science as a whole is an essentially anarchic enterprise. Feyerabend therefore contends that an examination of the history of science shows that there is only one 'principle' that does not inhibit progress:

> There is only *one* principle that can be defended under *all* circumstances and in *all* stages of human development. It is the principle: *anything goes.* (*AM¹*, p. 28, *AM³*, pp. 18–19)

This epistemological (better, *methodological*) 'anarchism' is asserted to be both more humanitarian and more conducive to scientific progress than its rationalist alternatives. Anarchism is the *only* advice which can be embodied in a *general* rule, which always helps to achieve progress. Of course, Feyerabend's apparent willingness to appeal to ideals such as 'progress', 'growth of knowledge', 'clarity', 'empirical success', 'discovery', 'advance' and 'improvement' in *AM* is misleading, since he disavowed his earlier, Popperian commitment to these ideals, declaring that *'Everyone can read the terms in his own way'* (*AM¹*, p. 27, *AM³*, p. 18).

In endorsing methodological anarchism Feyerabend is not suggesting a new and particularly liberal scientific methodology (although some critics misunderstood him thus). He explicitly declares that ' "anything goes" is not a "principle" I hold – I do not think that "principles" can be used and fruitfully discussed outside the concrete research situation they are supposed to affect – but the terrified exclamation of a rationalist who takes a closer look at history' (*AM*³, p. vii). He stresses not the absence but rather the *research-immanence* of methodological rules:

> The remarks made so far do not mean that research is arbitrary and unguided. There are standards, but they come from the research process itself, not from abstract views of rationality. (*SFS*, p. 99)

> [S]cience never obeys, and cannot be made to obey, stable and research-independent standards: scientific standards are subjected to the process of research just as scientific theories are subjected to that process; they do not guide the process from the outside. (*PP1*, p. xiii)

There is a strong connection here with the work of Michael Polanyi.[4]

Feyerabend characterizes the rules that methodological monists offer us as naïve and simple-minded (*AM*¹, p. 17, *AM*³, p. 9), strict and unchangeable (*AM*¹, p. 19, *AM*³, p. 11), firm and absolutely binding, (*AM*¹, p. 23, *AM*³, p. 14), fixed (*AM*¹, p. 27, *AM*³, p. 19), and as having been 'set up in advance and without regard to the ever-changing conditions of history' (*AM*¹, p. 18, *AM*³, p. 11). The kinds of rules he is thinking of are:

- Do not introduce *ad hoc* hypotheses in order to explain away recalcitrant data.
- Do not adopt hypotheses which contradict well-established and generally accepted experimental results.
- Do not adopt hypotheses whose empirical content is smaller than that of existing, empirically adequate hypotheses.
- Do not introduce self-inconsistent hypotheses.

Such rules, says Feyerabend, cannot 'account for' the historical complexity of science itself. So, taken as descriptions, they will not help observers to *understand* science. And, taken as prescriptions, they will be no good for participants, who ought to be opportunists rather than rule-followers.

Feyerabend's main argument for the epistemological anarchist conclusion was drawn from a long case-study of Galileo's defence of the heliocentric cosmology, a case-study grown from a seed planted in the mid-1950s by the logical positivist Philipp Frank:

Frank argued that the Aristotelian objections against Copernicus agreed
with empiricism, while Galileo's law of inertia did not. As in other cases,
this remark lay dormant in my mind for years; then it started festering. The
Galileo chapters of *Against Method* are a late result. (*KT*, p. 103)

In what one commentator entertainingly described as an 'irrational
reconstruction', Feyerabend tried to show that Galileo flouted every
methodological rule methodological monists have ever devised. The
power of his case is that it takes advantage of the cultural capital
invested in Galileo as one of the heroes of 'the scientific revolution'.
One widely shared basic value-judgement about our intellectual
history is that Galileo made progress over his predecessors. (Even the
Roman Catholic Church recently conceded that Galileo was, both
scientifically and morally, in the right.) AM is not supposed to be an
attack on Galileo, but rather an exhortation to us to remain faithful to
the methodological *radicalism* inherent in this basic value-judgement
on his behalf.[5] The importance of the scientific revolution, for
Feyerabend, is that it shows that truly radical conceptual change is
possible. Because it represents the sloughing-off of one conceptual
scheme, and the invention, defence, and triumph of another scheme
which had no logical links with the previous one, the scientific
revolution manifests our capacity for wholesale reconceptualization.
If Feyerabend is right, his case-study shows that this transition,
perhaps the single most important event in the history of human
thought, was and could only have been brought about by methods
which, to a methodological monist, are disreputable. And Feyerabend
goes on to claim that several other episodes would serve just as well
as the basis for similar case-studies, so that the scientific revolution is,
in this respect, typical of major conceptual changes in science.

We cannot go into the Galileo case-study here. It has been the
subject of some scrutiny by historians and philosophers of science.[6]
Many have noted that a single such case-study seems eminently
unsuited to make a general case against method. Methodological
monists could always reply by writing off Galileo as a charlatan, and
insisting that good scientists follow rules. But, as a matter of fact,
those who defend the idea of a single, fixed scientific method have *not*
been content to abandon Galileo. Instead, they have attacked
Feyerabend's interpretation of the case, arguing that he misidentified
Galileo's strategies, as well as the methodological rules Galileo really
endorsed. And perhaps the weakest part of Feyerabend's case is his
claim that the scientific revolution could *only* have been brought
about via 'irrational' means.

9.3 Deductivism and Inductive Methodological Rules

Let's focus instead on the general form of Feyerabend's argument. The methodological monist, who believes in 'firm, unchanging, and absolutely binding principles for conducting the business of science' (AM^1, p. 23, AM^3, p. 14) is *not* a straw man: Popperians and Lakatosians still subscribe to this view.[7] But the danger is that Feyerabend's critique fails to address those committed to the existence of inductive rules, and thus applies only to Popperians. Popperians commit only to rules which (in a certain sense) take no risks, rules which are deductive (Popper's 'deductive method of testing'). Seen in this light, their claim to be fallibilists is half-baked: they forswear the infallibilist aim of 'knowing for certain', but cling to an infallibilist and residually Cartesian ideal of scientific method in accepting only inferences whose conclusions can be drawn with certainty. The Popperian commitment to methodological monism and deductivism is thus a form of rule-fetishism honouring an ideal of method Popperians claim to have transcended. The idea behind it is that although 'knowledge' itself is not infallible, or even justified, and although the method might generate beliefs which are not knowledge, at least it cannot go wrong in allowing us to discard a theory that is not false. It is because Feyerabend is a Popperian deductivist *manqué* that he sees methodological anarchism as the only (other) option. Like Popper, he cannot accept that any rule could really work unless it is made exceptionless.

Only because of his Popperian ancestry does Feyerabend think he can refute methodological monism by demonstrating that methodological rules have been profitably violated. Like Popper, Feyerabend is at heart a deductivist. That is, he refuses to allow the possibility of inductive justification, and therefore searches for exceptionless deductive rules of methodology. Having criticized 'critical rationalism' for failing to realize that there are no such deductive rules, Feyerabend announces that there are no methodological rules at all. This strategy is obviously bankrupt. It fails to address any other 'rationalist' position. Inductive rules are perfectly good rules, and can be followed, can guide action towards a goal. Feyerabend's Popperian premises make it deceptively easy for him to reach his 'anarchist' conclusion.

Another way of putting this is to say that at the bottom of the weakness of Feyerabend's argument for methodological anarchism lies his inductive scepticism. Just as epistemological scepticism can be an over-reaction to the failure of infallibilism, so Feyerabend's extreme inductive scepticism is an infallibilist over-reaction to the problem of induction.

Could the kind of case-histories of profitable rule-violation which Feyerabend adduces suffice to refute the idea that there are norms of scientific methodology? Bill Newton-Smith, for one, thinks not, pointing out that wise methodological monists will endorse inductive rules, rules which advise us which of a pair of theories to adopt in the face of available evidence. But inductive rules are rules which we should expect to lead us astray on some occasions:

> [W]e should expect our rules to have a high risk factor. If our rules are too safe ... they may cushion us from error at the cost of minimising the number of contexts in which we actually end up adopting a theory. Thus, to have evidence of a number of occasions in which some rule has led us astray is not necessarily to have an adequate reason for doubting the acceptability of the rule. It may be, these exceptions notwithstanding, that our chance of progress in the long run is greater if we employ the rule. (Newton-Smith [1981], p. 129)

There are two stages in the refutation of methodological monism: certain information about the history of science would suffice to refute the naive falsificationist supposition that there is a single set of non-inductive methodological rules (Popper, as we shall soon see, seems to have got this far). Other methodological monists (for example Lakatosians) can allow that methodological rules are not exceptionless, since exceptionless rules would (as Feyerabend once thought) lead to no science at all, or at least would not lead to *our* science (as he later thought). But they cannot allow that methodology changes, since without a fixed methodology, relativism results. Simply because scientific methodology is a human creation, we ought to expect it to change over time.[8] But to methodological monists it is by no means obvious that a fixed method or theory of rationality did not spring into the world fully formed. Thus the second stage in the refutation of methodological monism necessitates extensive historical investigation in order to establish that there is no single set of inductive methodological rules. Newton-Smith is correct in suggesting that Feyerabend does not succeed in showing this. If Feyerabend's lesson is merely that theory-choice is less algorithmic than had previously been thought, then it is a salutary one. But this does not establish the non-existence, or even the plurality, of methodology.

Some considerations suggest that it might be possible to show that there is no set of inductive methodological rules followed by all good scientists. For the methodological monist to make her case, it is insufficient for her to find rules which *would* make sense of scientists' activities. Instead, she must find methodological rules which all good scientists have actually followed. But in order to count as really following a rule (and not just as conforming to it), a person must be credited with an understanding of that rule, an ability to explain it, and a willingness to justify behaviour by reference to it. This puts an upper bound on the complexity and obscurity of any methodological rules we can pin on rule-followers. It also means that such rules must be framed in terms of concepts that the rule-followers in question (the scientists) would have understood. It is by no means an easy task to show that there has been an invariant core of methodological rules which satisfy these requirements. Methodological monists, after all, conspicuously disagree as to exactly which rules good scientists followed. Scientists cannot be supposed to have been following rules which they (or, on their behalf, philosophers) are incapable of articulating. But as the rules specified by the 'rationalist' become more and more involved or recondite, our confidence that these were the rules that past scientists were really following ought to wane.

In the meantime, rational models of science have gone decidedly out of fashion within philosophy of science. This is partly a result of a general turn towards naturalistic epistemology. The latest attempts by philosophers to get to grips with the history of science are explicitly naturalistic, and there seems to be a growing awareness that the project of giving a rational model of scientific development has been a degenerating research programme.[9] It would be nice to think that this is partly due to Feyerabend's influence.

Methodological pluralism does not mean that science proceeds without reference to any rules or norms at all. It just means that science, even the best science, does not proceed in accordance with the same set of rules. The rules that constitute scientific method are not the same from science to science. The idea that theoretical physicists, experimental physicists, biologists, neurophysiologists, palaeontologists, ecologists, etc., all use the same method is incredible. Methodological monists have blinded themselves to this by drawing their examples largely from the physics of the last few hundred years. The methodological monist's dilemma now emerges: if she specifies a set of rules which are too loose, the set will not count as a distinctively scientific method. The only rules common to all sciences are global maxims like 'consider the evidence', 'test your theory', etc., but these do not comprise a methodology which is distinctive of science. But if she avoids this pitfall, she will be in danger of specifying a set of rules too tight to fit our pre-theoretical or even our informed judgements as to what counts as science. At any level of description at which all scientists can be described as using the same method, the method in question will not be distinctively scientific: it will fail to solve the 'problem of demarcation'.

Methodological monism is the very core of 'rationalism' because, for the monist, without a fixed methodology the idea of explaining the development of science as a rational process has been abandoned:

> If no principles of evaluation are fixed, then there is no 'objective view-point' from which we can show that progress has occurred and we can say only that progress has occurred *relative to the standards that we happen to accept now*. (Worrall [1988], p. 274)

The rationalist verdict is that any retreat from methodological monism amounts to an endorsement of relativism.[10] When the methodological pluralist draws attention to changes in the methodology subscribed to by scientists, the monist counters with the claim that there is a narrow sense of the term 'methodology' which picks out a core of principles that have *not* varied. For the monist, the examples of methodological change that pluralists come up with are inessential modifications which betoken invariance at a deeper level. The debate then threatens to bog down around a dispute over which sense of the term 'methodology' really captures the concept of methodology.

9.4   Popper's Methodological Anarchism

Although he berates Popperians for trying to take credit for absolutely everything,[11] Feyerabend points out that Popper himself anticipated methodological anarchism, when he reports that in his 1952 lectures at the LSE, Popper

> started with a line that became widely known: 'I am a Professor of Scientific Method – but I have a problem: there is no scientific method.' 'However,' Popper continued, 'there are some simple rules of thumb, and they are quite helpful.' (*KT*, p. 88)

'Anything goes' is, Feyerabend admits, an obvious practical consequence of such a 'critical rationalism' (*PP2*, p. 21).

It is not clear *how* Popper arrived at the conclusion that there is no scientific method. (And this is a conclusion, moreover, which neither he nor his followers advertise very extensively! Perhaps Popper meant only that there is no 'logic' of scientific discovery? If so, this is a misleading way of putting the point.) But he apparently thought that because there is no such thing, no set of rules which really explain how successful scientists proceed, we are obliged to move from a descriptive to a normative epistemology, a 'methodology' which would lay down, rather than discover, rules which scientists ought to follow. Feyerabend, in his early work, also committed himself to normative epistemology (although possibly not for the same reasons) and to falsificationism:

> Popper knew that a guide, or a map, may be simple, coherent, 'rational', and yet may not be about anything. Like Kraft, Reichenbach, and Herschel before him, he therefore distinguished between the practice of science and standards of scientific excellence and asserted that epistemology dealt only with the latter: the world (of science, and of knowledge in general) must be adapted to the map, not the other way around. For a while I reasoned in the same way. (*KT*, p. 90)

In this respect, Feyerabend was one of the strongest believers in a single codifiable scientific method. Then, in the mid-1960s, and probably as the result of studies in the history of science, he rediscovered Popper's methodological anarchist conclusion. Feyerabend actually gave credit for his conversion to methodological anarchism to the physicist C. F. von Weizsäcker, who, in a discussion in 1965, made him realize the futility of attempts to find general rules to cover all scientific moves.[12] In his autobiography, Feyerabend compares his own plea for theoretical pluralism with the more historical approach of Kuhn and Weizsäcker:

> Compared with this rich pattern of facts, principles, explanations, frustrations, new explanations, analogies, predictions, etc., etc., my plea seemed thin and insubstantial. It was well enough argued, but the arguments occurred in outer space, as it were; they had no connection with scientific practice. For the first time I felt, I did not merely think about, the poverty of abstract philosophical reasoning. (*KT*, p. 141)

Because of his growing dissatisfaction with normative epistemology (witnessed in his later suppression of passages from his early work), Feyerabend embraced the methodological anarchism Popper had swerved to avoid. But even after his conversion Feyerabend's complaint about the approach taken by Kraft, Popper, and his own former self is only that it exemplifies 'the dangers of abstract reasoning' (*KT*, p. 89), rather than the misguidedness of normative epistemology. Nevertheless I submit that methodological anarchism represents Feyerabend's belated recognition of the poverty of the normative approach to philosophy of science he learnt from Popper and Kraft.

Perhaps the root disagreement between Popper and Feyerabend, even the Feyerabend of *AM*, concerns only the idea that it would be *helpful* if scientists followed the simple falsificationist rules of thumb. The later Feyerabend's central complaint against Popper concerns simplicity: one of the most prominent themes of all Feyerabend's later work is that of the sheer *complexity* of science:

> [S]cience is a complex and heterogeneous historical process which contains vague and incoherent anticipations of future ideologies side by side with highly sophisticated theoretical systems and ancient and petrified systems of thought. (*AM*[1], p. 146)

Popper makes the mistake of trying to introduce simple rules of thumb, where only complex and context-dependent rules operate (the complexity of science being irrelevant to the normative epistemologist). It is not that Popper's rules would destroy any science ('Popper's rules can produce a Byzantine science; they are not entirely without results' (*KT*, p. 91)), but that they would destroy science *as we know it*. The most plausible parts of Feyerabend's conclusion are that the attempt to capture the world in explicit theories (scientific realism), and the attempt to capture science in simple rules (normative epistemology) are both misguided. This in no way means that we can draw the conclusions (which have been associated with epistemological anarchism) that no substantive theory is superior to any other, that one may defend any theory one likes, but that one has no right positively to assert or oppose any thesis, or that one must oppose ideals like truth, reason, knowledge and rationality.

For history to tell us that there is a single scientific method it is not sufficient that we find some rule or rules that all scientists followed. The rules discovered must be more than merely logical rules or common-sense rules of thumb if they are to count as rules of *scientific* method, not just rules of rational inquiry. If the conclusion of historical studies is that there is no single scientific method, we cannot take refuge in the thought that methodology is normative rather than descriptive. Normative epistemology, I have argued, is bankrupt. Must we then become epistemological 'anarchists'? It rather depends what this means. If it means only that we must recognize that science is not a natural kind, that 'science' is a family-resemblance term, that there is no single substantive method which marks off science from other activities (the view I have called 'methodological pluralism'), we may indeed be forced to accept it. Some philosophers and historians of science today seem to take this for granted:

[S]cientists have always acted in a loose and rather opportunistic way when *doing* research, though they have often spoken differently when *pontificating* about it. By now this has become a commonplace among historians of science. . . . I am far from claiming that the historians engaged in these new types of research have necessarily read *AM* and were educated by it – nothing would be further from the truth. But it is pleasant to see that some armchair views of mine are being held by scholars working in close contact with scientific practice. (*KT*, p. 151)

There is a certain strand of methodological pluralism in contemporary philosophy of science which can be traced back to Feyerabend, and which manifests scepticism about the alleged 'unity' of science.[13] According to this view, we should not exaggerate the uniformity of the sciences: different sciences simultaneously have very different methods, and stages in the history of a single science exhibit different methods too. This perspective results in a pluralism not about theories, but about the nature of science and knowledge:

Science and common sense are not as simple, self-contained, and faultless as the critics of their philosophical superstructures, myself included, were assuming. There is not one common sense, there are many . . . Nor is there one way of knowing, science; there are many such ways . . . Science itself has conflicting parts with different strategies, results, metaphysical embroideries. It is a collage, not a system. (*KT*, pp. 142–3)

Seen in this light, Feyerabend heralds the demise of the positivist 'Unity of Science' project, the attempt to specify either a single scientific method, or a relationship which theories from different branches of science must bear to one another in virtue of which their subject matters can be reduced to, or identified with, one another. This is somewhat ironic, given Feyerabend's strong initial commitment to the ideal of unity, and the role it plays in his argument for scientific realism.

Feyerabend's work does, however, contain an important warning for those attracted towards a naturalistic approach to scientific method. Such an approach urges us to look both at what scientists say, and at what they do. This is in line with the unimpeachable general advice that if one wants to know what rule a person is following, that person's sincere testimony is authoritative. Feyerabend's warning is that these two criteria, what scientists say, and what they do, often point in different directions. He argues that the classical physics of Galileo, Newton, Faraday and Maxwell exhibits split consciousness: the official methodology does not correspond to the actual scientific practice:

The official philosophy is still empiricism – and indeed a very militant empiricism it is. Speculation is discouraged; hypotheses are frowned upon; experimentation and derivation from observational results are regarded as the only legitimate manner of obtaining knowledge . . . We know now, mainly through the work of Duhem and Einstein, that none of these alleged derivations is valid; that the theories go *beyond* existing observational results and that they are also *inconsistent* with them. We therefore witness here the astonishing spectacle of men who invent bold new theories; who believe that these theories are nothing but a reflection of observable facts; who support this belief by a procedure which is apparently a deduction from observations; who in this way deceive both themselves and their contemporaries and make them think that the empirical philosophy has been strictly adhered to. We have here a period of *schizophrenia* characterised by a complete break between philosophical theory and scientific practice. This is an age when the scientist does one thing and insists that he is doing, and must do, another. ([1962c], pp. 138–9)

More generally, 'great scientists, while intuitively adopting a methodological opportunism, or anarchism of this kind almost always act as if they had followed a specific and well-defined method' ([1970e], p. 67).

Certainly, this is a problem that advocates of a naturalistic conception must be aware of. But it is not insuperable. After all, Feyerabend himself manages, somehow, to identify the real, underlying methodology of these scientists.

9.5   The Linguistic Relativity Principle

We have already noted (in chapter 6) that Feyerabend's argument for incommensurability works only if we grant him his peculiar version of scientific realism. That some theories are incommensurable when interpreted realistically, but commensurable when given an instrumentalist interpretation is reiterated in *AM* (see AM^1, pp. 274, 276, 278–82). If Feyerabend wants to establish the existence of incommensurability, he still has to have an argument for his version of realism. (Other, more familiar versions will not be effective substitutes, since they do not insist that observation-statements are interpreted in terms of theories.)

Feyerabend's admission that theories are incommensurable only under a realistic interpretation should bring us to realize the problematic nature of his realism. A realistic interpretation, for him, is simply one which ignores the consequences of the theory (more precisely, of the theoretical system) at the empirical level, focusing entirely on the

objects allegedly designated by its theoretical terms. We are supposed to think of Newton's theory, for example, as postulating a
world of wholly Newtonian objects (three-dimensional bodies in
space and time), and of relativity theory as postulating a world of
wholly relativistic objects (four-dimensional space-time 'worms').
The incommensurability thesis is now the view that theories interpreted in this way cannot be semantically compared. This seems
hardly surprising:

> What Feyerabend says on this point seems in the end to come dangerously
> close to the following triviality: We may say that Einstein's and Newton's
> theories make (possibly conflicting) statements about the same objects, but
> this is to interpret the two theories instrumentally. If we interpret it
> realistically then Newton's theory says something only about Newtonian
> objects (which, among other things, have, when involved in no physical
> interactions, constant spatial dimensions). It is thus entirely incommensu-
> rable with Einstein's theory which (when interpreted realistically) says
> something only about the entirely different class of Einsteinian objects
> (which, amongst other things, have velocity-dependent spatial dimen-
> sions). (Worrall [1978a], p. 292)

If what we mean by saying that two theories postulate the same
objects is that those theories are comparable at the empirical level, and
if what we mean by interpreting a theory 'realistically' involves
seeing it as postulating its own world of unique and incomparable
objects, the incommensurability thesis does indeed become a trivial-
ity. But Feyerabend should not endorse these equations. The problem
is that, in order to avoid the unpalatable conclusion that *any* two
theories are incommensurable when interpreted realistically, he will
be thrown back into the hazardous attempt to give criteria for when
theoretical entities should count as the same, and when they differ.

Feyerabend's admission that incommensurability is relative to a
realistic interpretation will clearly leave 'rationalists' unmoved: they
will simply eschew his version of scientific realism (safe in the
knowledge that this does not mean relinquishing any other form of
that view, or endorsing instrumentalism).

With the demise of his confidence in normative epistemology, and
thus in the 'principle of realism', we should expect to find Feyerabend
recognizing that his case for incommensurability is under threat. This
is not clearly so. In the first edition of *AM*, he presents the incommensurability thesis again, and the core of the first versions of the thesis
remains intact. Admittedly, it is here cast in a significantly different
light: not as the consequence of a theory of meaning, but as an
independent discovery, a result of applying what Feyerabend calls

'the anthropological method'. The strategy is to convince us that the phenomenon of incommensurability is identifiable via anthropological spadework, thus obviating the prior, philosophical question of whether such a phenomenon really makes sense at all.[14]

There can be little doubt that we are surreptitiously being sold a theory in this chapter of *AM*, not just having our attention drawn to a brute phenomenon. In his attempt to persuade us to the contrary, using what he supposes to be the anthropological method, Feyerabend appeals to two anthropologists from rather different traditions. The first is Benjamin Lee Whorf, who offered a theory of language centred on what is now known as the 'linguistic relativity principle' (or the 'Sapir–Whorf hypothesis'). This says that linguistic patterns determine both what the individual perceives and how that individual thinks about what he or she perceives. According to Whorf, the individual is not free to give an impartial description of nature. Rather he is

> constrained to certain modes of interpretation even while he thinks himself most free . . . We are thus introduced to a new principle of relativity, which holds that all observers are not led by the same physical evidence to the same picture of the universe, unless their linguistic backgrounds are similar, or can in some way be calibrated. (Whorf [1956], p. 214)

This linguistic relativity principle (supposedly modelled on the physicist's 'principle of relativity') means, Whorf argues, that

> users of markedly different grammars are pointed by their grammars toward different types of observations and different evaluations of externally similar acts of observation, and hence are not equivalent as observers but must arrive at some different views of the world. (p. 221)

Feyerabend mentions that Whorf moves between two different interpretations of this claim, without clearly distinguishing between them. The apparently less radical doctrine is to the effect that observers using widely different languages will *arrange similar facts in different ways* (*AM*[1], p. 286), that is, that they will 'report' or 'classify' the same experience in different terms. This is a version of what we might call 'conceptual relativism', according to which different observers can have different conceptual schemes. The apparently more radical thesis, and the one which Feyerabend is tempted by, is that such observers will *posit different facts* under the same physical circumstances in the same physical world' (*AM*[1], p. 286). This he calls an 'extension' of Whorf's view. But in fact this view is *less* radical than Whorf's conceptual relativism: it may be that different observers who 'posit different facts' under the same physical circumstances are responding to completely different features of their environment. For example, observers with different sense-organs may posit different facts under the same circumstances simply because they are responding to data from different parts of the electromagnetic spectrum. This does not mean they do not share a conceptual scheme.

Feyerabend also deploys the concept of a 'covert classification', which Whorf defines thus:

> A linguistic classification like English gender, which has no overt mark actualized along with the words of the class but which operates through an invisible 'central exchange' of linkage bonds in such a way as to determine certain other words which mark the class, I call a *covert class*, in contrast to an *overt* class, such as gender in Latin. (p. 69)

These 'linkage-bonds', in the case of English gender, connect gender nouns ('boy', 'girl', 'wife', Christian names, etc.) either to the word 'he' or to the word 'she'. A covert linguistic class, Whorf says, 'may have a very subtle meaning, and it may have no overt mark other than certain distinctive "reactances" with certain overtly marked forms' (p. 70).

According to Feyerabend these 'covert classifications' have two functions. They underpin the cosmology which the grammar of a language 'contains' (a cosmology being 'a comprehensive view of the world, of society, of the situation of man which influences thought, behaviour, perception' (AM^1, p. 223)). And they create 'patterned resistances' to greatly divergent points of view, resistances which 'oppose not just the truth of the resisted alternatives but the presumption that an alternative has been presented' (AM^1, p. 224). In such a case, Feyerabend tells us, we have an instance of incommensurability. But because incommensurability 'depends on covert classifications and involves major conceptual changes', 'it is hardly ever possible to give an *explicit definition* of it . . . The phenomenon must be *shown*' (AM^1, p. 225; emphasis added).

Feyerabend can only accept the linguistic relativity principle as applied to thought and behaviour since, as we already mentioned, he rejects the thesis that languages or theories determine perceptions themselves. But the central problem with any appeal to this principle is that insofar as it can be made testable it cannot be said to be well confirmed. Although there is no clear consensus, much of the immense literature on 'universals of human thought' bears negatively on the Sapir–Whorf hypothesis.[15] There also seems to be a problem of self-refutation lurking: how does Whorf himself escape his own stricture that none of us can describe nature with absolute impartiality?

Much of *AM*'s chapter 17 is devoted to showing us instances of incommensurability. There are examples from the field of perception, where different mental 'sets' prevent us from attending to and simultaneously comparing the two incommensurable aspects of familiar gestalt figures. There are examples from developmental psychology, where the 'visual world' of the child proceeds through a succession of stages each of which destroys the previous stage and leads to a reorientation of behavioural patterns and thought. In such a developmental process, we are told,

> we may suspect that the family of concepts centring upon 'material objects' and the family of concepts centring upon 'pseudo-after-image' are incommensurable in precisely the sense that is at issue here; these families cannot be used simultaneously and neither logical nor perceptual connections can be established between them. (*AM*[1], pp. 228–9)

Lastly, we are treated to an example from the history of art, in which the 'archaic style' of the ancient Greeks, and its underlying cosmology, are contrasted with a later cosmology initiated by the pre-Socratics.[16] The motive for this investigation is the drive to 'break through the boundaries of a given conceptual system, and to escape the reach of "Popperian spectacles"' (*AM*[1], p. 229). It involves the ability to '*produce* and *grasp* new perceptual and conceptual relations' including 'covert classifications' which are *ex hypothesi* hidden from view. We are enjoined not to forget, as 'orthodox accounts' do, that such covert relations contribute to the meaning of theories. Orthodox accounts are now placed in doubt, the concepts of observation, test, theory and truth rendered suspect. Therefore, when investigating and presenting incommensurability one should examine means of representation other than languages or theories, and styles of painting are well suited to this purpose.

The resulting account of incommensurability, which Feyerabend describes as an attempt to 'find terminology for describing certain complex historical anthropological phenomena which are only imperfectly understood rather than defining properties of logical systems that are specified in detail' (*AM*[1], p. 269), is as follows. Sufficiently general and comprehensive points of view, 'theories', are built in accordance with different principles of construction. Such principles are open to violation (which does not amount to contradiction) by certain statements. The production of such a statement suspends the principle which is violated. It does so by making inapposite the

'grammatical habit' which corresponds to the principle. Some principles underlie *every* element of the theory, they are the 'universal
principles'. When the universal principles of a theory are suspended
all facts accounted for by the theory and all concepts of the theory are
likewise suspended. Two theories are *incommensurable* if one suspends some of the universal principles of the other.

This new formulation does nothing to bring the debate over the
intelligibility of incommensurability to an end. We are not really told
how to identify and individuate 'principles of construction', 'universal
principles' or 'grammatical habits', or what it is for such principles to
be 'violated' or 'suspended' by any other principle or statement.
Feyerabend rails against those who would force him into clarifying the
terms of this characterization, but his tirade rests partly upon the very
worst part of *AM*, an indefensible relativism with respect to logic.

The text makes it clear that Feyerabend is struggling to present
what he thinks of as a non-linguistic concept in linguistic garb.
Although his account uses sentential terminology like 'principles',
these terms 'are supposed to summarise anthropological information'. Feyerabendians might see this as a sign of the paucity of a
linguistic approach to theories, and as an invitation to pursue a non-
sentential, structural conception of human knowledge.[17] Feyerabend
does not, however, withdraw his assertion that incommensurability
can be recognized only by those who have taken the decision to be
theoretical realists.

9.6 The Anthropological Method

Feyerabend insists that the existence of incommensurability is a
'problem of historical fact' and not of logical possibility. This blunt
dismissal of so many recent philosophical reservations is motivated
by his adherence to what he calls 'the anthropological method'. On
behalf of this he makes some extravagant claims:

> My argument presupposes, of course, that the anthropological method is
> the correct method for studying the structure of science (and, for that
> matter, of any other form of life). (*AM*[1], p. 252. See also [1977a], p. 363)

> Here as elsewhere only the anthropological method can lead to knowledge
> that is more than a reflection of wishful thinking. (*AM*[1], p. 266 n. 120)

> We need the anthropological method to find out whether reconstruction
> improves science, or whether it turns it into a useless though perfect

adornment of logic books. The procedure of the anthropologist therefore takes preference over the procedure of the logician. ([1976]: *PP2*, p. 204)

The only anthropologist mentioned when Feyerabend appeals to 'the anthropological method' is E. E. Evans-Pritchard, perhaps the most illustrious member of an influential British school of thought in that discipline.

The method presented by Feyerabend has three key elements, or stages. First, the anthropologist must learn the language and the basic social habits of the group he intends to study. He must inquire how these are related to other activities, some of which may seem of little importance. This process involves the identification of key ideas, and the attention to (*prima facie* insignificant) minutiae.

Second, having found these key ideas, the anthropologist must then try to understand them:

> This he does in the same way in which he originally gained an understanding of his own language . . . He *internalises* the ideas so that their connections are firmly engraved in his memory and his reactions, and can be produced at will . . . *This process must be kept free from external interference.* For example, the researcher must not try to get a better hold on the ideas of the tribe by likening them to ideas he already knows, or finds more comprehensible or more precise. On no account must he attempt a 'logical reconstruction'. Such a procedure would tie him to the known, or to what is preferred by certain groups, and would forever prevent him from grasping the unknown ideology he is examining. (*AM¹*, p. 250)

In the process of clarification which determines the content of the native concepts, the researcher is exhorted to restrain his urge for logical perfection. He should never try to make a concept clearer than is suggested by the material (except as a temporary aid to further research), since this material, not his logical intuition, determines the content of the concepts. Evans-Pritchard was certainly an advocate of this kind of hermeneutic understanding, which social thinkers and philosophers call 'verstehen'. However, the role, efficacy, and nature of verstehen have often been disputed.[18]

The third element of the procedure is the comparison of the group's society and cosmology with those of the anthropologist himself. Such a comparison 'decides whether the native way of thinking can be reproduced in European terms . . . or whether it has a "logic" of its own, not found in any western language' (*AM¹*, pp. 250–1). Any expression of the group's ideas in the European language is evidence only that languages are flexible, not that the two languages in question are truly commensurable.

What is noteworthy about this three-stage method is that *transla-tion* enters only at the third stage, and then only as a hindrance, rather than an aid, to understanding. Feyerabend repeatedly insists that the researcher must learn the alien language by 'immersion', as a child learns a language, and not by the study of abstract principles. Through-out his later work he ridicules those who would believe that such a process is impossible. The 'guiding idea' of these objectors, that a new language cannot be introduced directly, but must first be connected with an existing observation-language is, he complains,

> refuted at once by noting the way in which children learn to speak – they certainly do not start from an innate observation language – and [by] the way in which anthropologists and linguists learn the unknown language of a newly-discovered tribe. (*AM*[1], p. 280)

Is it to be assumed, we are asked, that grown-up philosophers and field researchers cannot perform the same feats of understanding which little children engage in every day?

The irony of appealing to a single correct anthropological method in a book whose central thesis is that there is no single scientific method has not gone unnoticed. But the appeal is not just ironic, it does no justice to the diversity of anthropological traditions and the centrality of debate over method within anthropology itself. Feyerabend's three stages constitute a reasonable elaboration of Evans-Pritchard's later conception of the anthropological method.[19] But Feyerabend goes beyond Evans-Pritchard in his account of lan-guage-learning. Evans-Pritchard of course insists that the anthro-pologist must learn the language of the people studied, and that he or she must speak only that language in the field. He even says that a complete understanding of the natives' language amounts to an understanding of their society. But he does not pronounce on whether this involves co-ordinating the natives' language with one's first language, or whether one must learn the new language 'as a child', from scratch.

Feyerabend's insistence that one can and should learn a new language without connecting it to any language one already speaks raises certain problems. If I understand a language A, then I cannot be credited with an understanding of another language, B, unless I can generally perform translations from B into A and vice versa. (Chil-dren who are bilingual but have not grasped the concept of transla-tion are not really exceptions to this principle.) Being able to translate from one's second language into one's first language is a *criterion* of understanding a second language. (The fact that there are other criteria of language-understanding does not affect this point).

Two rather different sets of considerations have been taken to suggest a positive answer to Feyerabend's rhetorical question about the impossibility of learning a second language 'from scratch'. One such set, which has also cast further serious doubt on the *intelligibility* of incommensurability, concerns the conditions under which language-learning is possible. Contemporary philosophers of language such as Quine and Davidson have made a strong case for the conclusion that translation of an alien tongue can only proceed on the basis of the assumption that the aliens' central beliefs do not depart drastically from our own. The argument is somewhat involved, but we can sum up its crux as follows. Imagine that you are trying to understand an alien whose language you do not speak. In trying to translate her words, you are forced to guess what they mean. Your guesses will inevitably start with words which are used in situations where perceivable material things are perceived by you both. And your guesses can only proceed on the assumption that your alien informant knows roughly what she is talking about. (If you've happened upon the village idiot, your project is in trouble.) But when you try to make sense of what your informant says, all you have to go on are your own opinions as to what makes sense. Trying to maximize the intelligibility of what the informant says therefore means trying to maximize the *truth* of what she says (this constraint is known as the 'principle of charity'). This means that you cannot seriously impute to the aliens basic beliefs which differ wildly from your own. If, having got some way with your translation, you find that the very next alien utterance translates into something that you consider to be not only obviously false but also destructive of a large set of other beliefs, you *must* suspect that your translation, rather than their belief, is at fault. If you blithely stick to your translation-hypothesis under these circumstances, you are, in a very real sense, not doing the aliens the justice of crediting them with sensible beliefs. So the supposition that they might have a very different 'conceptual scheme' goes by the board.

So far Quine. Davidson goes further, seeking to call into doubt 'the very idea of a conceptual scheme', an idea which seems essential to the presentation of any case for incommensurability (Davidson [1973]). Feyerabend, it must be said, made no adequate response to these arguments.[20] But there are replies that can be made on his behalf. The most powerful is perhaps simply that while the 'principle of charity' seems mandatory when translating very basic everyday beliefs, it seems wholly gratuitous when it comes to translating the theoretical beliefs that Feyerabend's incommensurability-thesis is concerned with.[21]

The other set of considerations which cast doubt on Feyerabend's view of language-learning is empirical. Whether human beings can learn a second language in the same way as they learnt their first is an empirical question. Learning a second language with no overt reference to one's mother tongue is known as the 'direct method' (as opposed to the 'grammar–translation method'). But while this is certainly possible, it does not establish that the two languages are not being co-ordinated by the learner in any way. A significant proportion of the deviant utterances produced by co-ordinate bilinguals (those who keep their two languages 'in separate compartments', as it were) can be attributed to interference from the mother-tongue.

There is also a raft of empirical considerations which suggest that learning a second language without relating it to one's mother-tongue in any way is impossible. Some studies of the optimal age for second-language learning suggest that the brain plasticity, biological predisposition and imprinting tendency required for language-learning all diminish with age. According to the first of these hypotheses, favoured by Wilder Penfield and Eric Lenneberg, there is a 'critical period' during which language acquisition can occur naturally and easily, but after which (around puberty) the brain is unable to process language input in this way.[22] Anthropologists, who fall outside of the critical period, must therefore learn second languages as grafts on their mother-tongue. The self-reflective musings of anthropologists, which Feyerabend appeals to as his evidence for the correct method in anthropology, are not going to be of great weight against such considerations: they may well be examples of scientists *mis*describing their own method. If this is so, Feyerabend himself has misrepresented 'the anthropological method'.

10

Relativism, Rationalism and a Free Society

10.1   Truth, and other Epistemic Ideals

Feyerabend is probably most famous for explicitly embracing the supposedly risqué view known as *relativism*. Since the turn of the century, relativism has been considered in almost every area of philosophy, and has penetrated far beyond philosophy into social science, literary criticism, and everyday thought. Because certain kinds of relativism have recently come in for much criticism within analytic philosophy, we owe it to self-confessed adherents to get clear exactly which kinds of relativism they endorse.

Relativism is often the result of a disenchantment with 'rationalism', a doctrine which relies upon the universal applicability of epistemic ideals (scepticism is another possible outcome). Before we see what kinds of relativism Feyerabend endorsed, it will be useful to take note of the development in his attitude towards epistemic ideals generally. As we have seen, Feyerabend originally, as a fairly orthodox Popperian, honoured and invoked classical epistemic ideals like truth, knowledge, rationality, honesty and scientific progress. His pluralistic test-model will not function without them. The first signs of his impending break with Popperian rationalism come in the early 1960s. His later disenchantment with the concept of truth may have its roots in his earlier conviction that truth is just too cheap in the 'myth predicament', the situation arising from the dominance of a single all-embracing theory (see chapter 5). In such a predicament, Feyerabend holds, it will seem that at last an absolute and irrevocable truth has been arrived at. Disagreement with facts may of course occur, but, being now convinced of the truth of the existing point of view, its proponents will try to save it with the help of *ad hoc* hypotheses. Experimental results that cannot be accommodated, even with the greatest ingenuity, will be put aside for later consideration. The result will be absolute truth, but, at the same time, it will decrease in empirical content to such an extent that all that remains will be no more than a verbal machinery which enables us to accompany any kind of event with noises (or written symbols) which are considered true statements by the theory. ([1962a]: *PP1*, p. 75)

It is interesting that here the result is said to be that the theory *is* absolutely true. Maybe we could write this off as a sloppy turn of phrase,[1] but it occurs more than once in writings of the same period. In the lectures entitled 'Knowledge without Foundations', for example, there is a fascinating discussion of the status of myths:

The power of a myth is not at all exhausted by the psychological factors we have described so far. A myth can very well stand on its own feet. *It can give explanations*, it can reply to criticism, it can give a satisfactory account even of events which *prima facie* seem to refute it. It can do this *because it is absolutely true*. It has, therefore, something to offer. It has to offer truth, absolute truth. ([1961a], p. 38)

The assertion that absolute truth is attainable through myth is premised upon an identification of absolute truth with total immunity from falsification. This is the kind of 'truth' which all-encompassing myths possess. Such an identification, coming from a close student of Popper's work, is both astounding and misconceived.

Later, when searching for a new ideal which would integrate the activities we now distinguish as arts and sciences, Feyerabend expressed scepticism about the old ideal of 'Truth':

This new ideal should be simply to make people more pleasant and more interesting, to make life happier, to make the world better, to make the beer better, and so on. These are all reasonable. Moreover, I am not being facetious when I mention things like beer. For I think it important to cut down to size those who say we must concern ourselves with nothing less than the search for Truth. For what Truth is nobody really knows. ([1968b], p. 130)

Feyerabend's disenchantment with traditional epistemic ideals began to surface in earnest in the early 1970s. Every mention of 'the Truth' begins to appear in scare quotes, or with a mocking capital letter.[2] Feyerabend declares that, in contrast to Lakatos, he does not intend to use 'the theological term *true*' ([1970b], p. 342 n. 105), and that 'truth' is a 'slogan' of Critical Rationalists which might be 'unimportant, and perhaps even undesirable?' ([1970c], p. 73).[3] By the time of *Against Method*, Truth (along with Reason, Honesty, and Justice) is an idol whose authority must be 'undercut' by epistemological anarchism (*AM*[1], pp. 32–3). The implementation of an anarchistic epistemology (even if only in a medicinal capacity) means that

> Reason, at last, joins all those other abstract monsters such as Obligation, Duty, Morality, Truth and their more concrete predecessors, the Gods, which were once used to intimidate man and restrict his free and happy development: it withers away. (*AM*[1], p. 180. See also pp. 189, 230)

In Feyerabend's first major discussion of rationalism, the very idea of truth was held to be a relatively recent product arising only with the first rationalists, the pre-Socratics. Although he did still use the concept of truth on occasion in his later work, he claimed to use it in an *ad hominem* capacity only, and certainly not in its capacity as a 'universal idea'.[4]

10.2   Rationalism vs. Relativism

We have already seen two possible routes to relativism within Feyerabend's work. First, methodological pluralism seems to entail a kind of relativism: if there is no fixed scientific method, changes in methodology cannot be justified by reference to a fixed core of rational principles.[5] There would then be no way of rationally assessing the comparative merits of different paradigms. Second, although a thin version of incommensurability can be accepted by non-relativists, incommensurability too, when combined with a suitable incomparability thesis, entails relativism. If incommensurability precludes our making theory-comparisons which are not based on truth-content, relativism ensues. Although, as we have seen, Feyerabend was accused of being a relativist a number of years before his conversion to methodological pluralism and the incomparability thesis, he did eventually embrace these views, and the relativism they entail, with gusto.

In this chapter, I shall concentrate on what Feyerabend actually says about relativism. To anyone familiar with the usual critical discussions of relativism, Feyerabend's rare comments on the subject will come as some surprise. His first sustained discussion occurs in a little-known paper of 1977, 'Rationalism, Relativism and Scientific Method'. Here he provides his own taxonomy of rationalisms, distinguishing between naive and sophisticated versions of 'cosmological', 'institutional' and 'normative' rationalisms respectively. Opposing all these forms of rationalism, Feyerabend presents the idea that there is not *one* rationality, there are *many* and it is up to us to choose the one we like most. This doctrine is 'Relativism':

> For many thinkers such a result is intolerable. Relativism, they believe, opens the door to chaos and arbitrariness. The fear of chaos, the longing for a world in which one need not make fundamental decisions but can always count on advice, has made rationalists act like frightened children. 'What shall we do?', 'How shall we choose?', they cry when presented with a set of alternatives assuming that the choice is not their own, but must be decided by standards that are (a) explicit and (b) not themselves subjected to a choice. Relativism, however, brings choice into everything – hence the aversion. ([1977a], p. 16)

The rationalist would undoubtedly reply that there is a false contrast implied here between choices which are 'one's own' and choices made according to standards or rules. In Feyerabend's thought rationalists appear as people who cannot make their minds up without the external intervention of such explicit rules. According to this picture, following a rule or conforming to a standard involves the deployment of some representation of that rule or standard in conscious thought. To this Feyerabend retorts that

> Traditions, institutions influence behaviour not only via rules and standards that can be made explicit. When recording an observation, reacting to a smile, checking the results of a complicated calculation, we act 'automatically', without consulting explicit rules and without being able to say what rules were involved. ([1977a], p. 16)

This point is well taken. Perhaps some philosophers have been implicated in the simplistic picture of rule-following which Feyerabend ascribes to the rationalist.

Underlying Feyerabend's dissatisfaction with rationalism, however, is a more powerful sceptical objection. Rationalists, because they object to certain ways of reasoning, insist that failures of rationality are *possible*. It seems that they must therefore concede that there is a choice of ways of reasoning, that it is possible to choose between rational and irrational belief-forming methods. However, this choice cannot possibly be a rational one. If it *were* rational, the chooser would already have committed himself or herself to rational standards, and their choice would not be an ultimate choice. Here we are therefore faced with what has been aptly called the 'dilemma of ultimate commitment': our ultimate commitment, our commitment to rationality or to (some form of) irrationality, can only be made nonrationally.[6] So rationalists and irrationalists are ultimately in the same predicament: each has made a (rationally unjustifiable and uncriticizable) leap of faith. A choice in favour of rationality must itself be a non-rational choice. Feyerabend uses this to present a dilemma to the rationalist, one horn of which is an infinite regress, the other one which has become familiar from the work of Wittgenstein and Polanyi:

> Assume we want to judge action A by standard S. We apply S to A and render our judgement. But the application must also be rational, so there must be standards S' which judge the pair {A,S} and so on *ad infinitum* unless we admit that at some place we simply act without being able to provide the standards which make this action rational. ([1977a], p. 16)[7]

This objection to rationalism represents Feyerabend's belated recognition of one of Wittgenstein's central points: justification comes to an end, but it comes to an end (at the bottom of the 'language-game') not in some more basic theory but in action, our natural and common human propensity to act thus rather than in some other way. Our basic beliefs or commitments are, in this sense, groundless.[8]

Feyerabend indicates that the relativist heroically dispenses with the alienated authority invested in explicit rules and standards, by striking out on his own: 'He knows that every step he takes is a step into darkness. He may end up in obscurity and empty verbiage; but he may also find new canons of action and understanding' ([1977a], p. 18). However, Feyerabend's account of the relativist path makes it sound more like an extreme form of voluntarism: breaking away from traditions is equated with being able to dispense with all rules; the act of 'choice', the choice between traditions, is revered as pure and ultimate. Most significantly, the article ends with the assertion that

> One might call the omnipresence of this choice the 'existential dimension' of research. The fact that there is such an existential dimension to every single action we carry out shows that rationalism is not an agency that forms an otherwise chaotic material, but is itself material to be formed by personal decisions. The questions 'What shall we do?' 'How shall we proceed?' 'What rules shall we adopt?' 'What standards are there to guide us?' however are answered by saying: 'You are grown up now, children, and so you have to find your own way'. (p. 19)

Voluntarism is, as I noted in chapter 1, a very strong theme throughout Feyerabend's thought.[9] It chimes with his insistence that ethical decisions dictate at least the form of human knowledge.

10.3 Idealism, Naturalism, Interactionism

More clues to the nature and extent of Feyerabend's relativism emerge in *Science in a Free Society*. Again the discussion occurs as part of a tirade against 'rationalism', and Feyerabend starts by introducing two opposed views about the nature of rationality: idealism and naturalism. Idealists treat practice as raw material to be manipulated by reason, the latter being conceived of as an objective, external agency with which we can shape our activities. On such a view, science is the result of our application of reason to some 'partly structured, partly amorphous material' (*SFS*, p. 7). Naturalists, by contrast, favour concrete practices over abstract reason. They suppose that any attempt to interfere with the relevant practices, practices which have their own kind of integrity, would result in their breakdown, not in their perfection. They do not trust the application of abstract reason, since its results are always at the mercy of prevailing conditions. Feyerabend (who, in these terms, started as an idealist but moved towards naturalism) plays these two views off against one another, and then tentatively puts forward a compromise position which he calls 'interactionism':

> Interactionism means that Reason and Practice enter history on equal terms. Reason is no longer an agency that directs other traditions, it is a tradition in its own right with as much (or as little) claim to the centre of the stage as any other tradition. Being a tradition it is neither good nor bad, it simply is. The same applies to all traditions – they are neither good nor bad, they simply are. They become good or bad (rational/irrational; pious/ impious; advanced/'primitive'; humanitarian/vicious; etc.) only when looked at from the point of view of some other tradition. 'Objectively' there is nothing to choose between anti-semitism and humanitarianism. But racism will appear vicious to a humanitarian while humanitarianism will appear vapid to a racist. *Relativism* (in the old and simple sense of Protagoras) gives an adequate account of the situation which thus emerges. (*SFS*, pp. 8–9)

The perspectival illusion referred to here is allegedly grounded in the fact that value-judgements are implicitly relational. Only when traditions are equally powerful is this important feature vividly apparent, for when one tradition exerts its power over a weaker rival it will 'educate' its victims into believing in its superiority. Such an illusion is the result of a pernicious conflation of the standpoints of the participant and the observer:

> When speaking as observers we often say that certain groups accept certain standards, or think highly of these standards, or want us to adopt these standards. Speaking as participants we equally often *use* the standards without any reference to their origin or to the wishes of those using them. We say 'theories ought to be falsifiable and contradiction free' and not 'I want theories to be falsifiable and contradiction free' or 'Scientists become very unhappy unless their theories are falsifiable and contradiction free'. Now it is quite correct that statements of the first kind (proposals, rules, standards) (a) contain no reference to the wishes of individual human beings or to the habits of a tribe and (b) cannot be derived from, or contradicted by, statements concerning such wishes, or habits, or any other facts. But that does not make them 'objective' and independent of traditions ... There are many statements that are *formulated* 'objectively' i.e. without reference to traditions or practices but are still *meant to be understood* in relation to a practice. Examples are dates, co-ordinates, statements concerning the value of a currency, statements of logic (after the discovery of alternative logics), statements of geometry (after the discovery of Non-Euclidean geometries) and so on. The fact that the retort to 'you ought to do X' can be 'that's what *you* think!' shows that the same is true of value statements. (*SFS*, pp. 22–3)

Here we have a crude version of value-relativism: all value-judgements are implicitly relational, and any such judgement not explicitly relativized to a tradition is 'fatally incomplete' (*SFS*, p. 24). Coupled with this old idea is Feyerabend's thesis that reason is but one tradition among others.[10]

From his discussion of the relation between reason and practice, and from his conclusion that '*What is called "reason" and "practice" are ... two different types of practice*' (*SFS*, p. 26) Feyerabend draws a series of characteristically relativist implications, the first few of which he announces as follows:

> i. *Traditions are neither good nor bad, they simply are.*

> ii. *A tradition assumes desirable or undesirable properties only when compared with some tradition* i.e. only when viewed by participants who see the world in terms of its values.

> iii. *i. and ii. imply a relativism of precisely the kind that seems to have been defended by Protagoras.* Protagorean relativism is *reasonable* because it pays attention to the pluralism of traditions and values. And it is *civilised* for it does not assume that one's own village and the strange customs it contains are the navel of the world. (*SFS*, pp. 27–8)

Feyerabend came to identify more and more with the views of Protagoras, the original relativist, whose position was criticized by Plato. Protagoras began his book on 'Truth' by saying that: 'Man is the measure of all things; of things that are that they are, and of things that are not that they are not.' This gnomic utterance can be construed in different ways. Perhaps the most charitable has mankind as its subject. On this interpretation, the thought expressed is a version of anthropocentrism, in which the relativity of judgement is to mankind itself. Neither Plato nor Feyerabend considers this interpretation, however, perhaps because it does not fit with enough of the other things Protagoras is reported to have said.

If, on the other hand, it is *individual people* referred to by the term 'man', as Plato and Feyerabend believe, then Protagoras' dictum may express a form of *subjective relativism*, according to which the truth of a judgement is relative to the maker of that judgement. This is a way of saying that for each person things have just those qualities which they seem to that person to have, but have no qualities (or perhaps: no existence?) independently of human observers. The problem with subjective relativism is that it collapses the distinction between belief and (relative) truth. Arguing against a person who genuinely believes something different consists in trying to show that what they believe is somehow inadequate. But faced with someone, A, who believes in absolute truth, the subjective relativist has to concede that absolutism is true-for-A, that because some people believe truth to be absolute it *is*, for them, absolute. Subjective relativism therefore cannot offer a *generally* valid account of the nature of truth.

Feyerabend's focus on social traditions and institutions actually suggests that it is not an individualistic view like this to which he really subscribes. This is confirmed in *Farewell to Reason*, where Feyerabend gives the following gloss on Protagoras' pronouncement: 'Whatever seems to somebody, is to him to whom it seems' (*FTR*, p. 45). This is the subjective relativism Plato attacked. Feyerabend concedes that Plato's counter-arguments were decisive, but proposes another interpretation of Protagoras' pronouncement which does not succumb to Platonic objections:

Laws, customs and facts . . . ultimately rest on the pronouncements, beliefs and perceptions of human beings, so important matters should be referred to the people concerned and not to 'abstract agencies and distant experts'. (*FTR*, p. 48)

This certainly captures something of Protagoras' philosophy (as introduced, for example, in Plato's dialogue *Theaetetus*). But it surely evinces a commitment to what Feyerabend called naturalism and to the political view he called 'democratic relativism' (for which, see section 10.4 below), rather than to relativism as such. What we were expecting from Feyerabend is a form of objective relativism, the view which insists that there can be objective reasons within a shared, public framework of presuppositions. Protagoras' dictum, though, does not easily lend itself to such a construal.[11]

Further light is shed on Feyerabend's views later in *Science in a Free Society* when he distinguishes between 'philosophical' and 'political' relativisms:

> [W]e must distinguish between political relativism and philosophical relativism and we must separate the psychological attitude of relativists from both. *Political relativism* affirms that all traditions have equal *rights*: the mere fact that some people have arranged their lives in accordance with a certain tradition suffices to provide this tradition with all the basic rights of the society in which it occurs . . . *Philosophical relativism* is the doctrine that all traditions, theories, ideas are equally true or equally false or, in an even more radical formulation, that any distribution of truth values over traditions is acceptable. (*SFS*, pp. 82–3)

But we are immediately informed that

> This form of relativism is nowhere defended in the present book. It is not asserted, for example, that Aristotle is as good as Einstein, it is asserted and argued that 'Aristotle is true' [*sic*] is a judgement that presupposes a certain tradition, it is a relational judgement that *may* change when the underlying tradition is changed . . . Value judgements are not 'objective' and cannot be used to push aside the 'subjective' opinions that emerge from different traditions. (*SFS*, p. 83)

So, rather disappointingly for those who would like him to nail his colours to the mast, Feyerabend identifies 'philosophical relativism' only to disassociate himself (or at least this book) from it.[12] It is interesting, however, that he explicitly classes the statement 'Aristotle is true' as a value-judgement. This strongly suggests that any statement expressing a comparative assessment of the merits of different scientific theories (that is, any comparative judgement of predictive power, explanatory power, empirical adequacy, verisimilitude, simplicity, etc.) will count as a value-judgement, and will thus be regarded as implicitly relational. Such a consequence would be an intelligible and predictable conclusion to draw from Feyerabend's later work. It is certainly worth bearing in mind that this too must be an integral part of what we mean by 'Feyerabend's Relativism'.

10.4   'Democratic Relativism'

When his views within philosophy of science demanded that he evaluate the social standing of science, the arguments for taking its theories and results as trustworthy, and for granting to it its current status, Feyerabend began to delve more deeply into social and political philosophy. The issue that particularly concerned him was the role of science within contemporary democratic society. The traffic between these institutions is two-way. The question of the relationship of science to society seemed important enough, to Feyerabend, to be able to generate answers to general questions in political philosophy; and, correlatively, he sometimes used a pre-established political philosophy in order to criticize the standing of science within society.

The main focus of his political philosophy and the keystone of his idea of a free society was what he originally referred to as 'political relativism', the view that all traditions have equal rights. Feyerabend took this view, which he soon came to call 'Democratic relativism', and which first raised its head at the end of *Against Method*, to be a political consequence of epistemological anarchism. He made the connection explicit in chapter 18 of *AM*[1] and in 'Democracy, Elitism, and Scientific Method', where he considered the question 'How should the citizens of a free and democratic society judge the institutions that surround them and the things they produce?'. These citizens must evaluate the effects and achievements of their society's institutions, in order to be able to change and control them. To do this, they need standards. Which standards should they use? Many will agree that they should choose rationally, and some will seek to identify rational choice with choice according to the standards of scientific rationality. The first difficulty with this proposal, says Feyerabend, is that there is no single set of unambiguous scientific standards, rather, there are many different proposed methodologies (Newton's, Mill's, Popper's, etc.), and science itself runs on a fruitful and opportunistic pluralism of ideas. But philosophy of science, and the more homogeneous, abstract, and historically irrelevant conception of reason which it embodies, have deteriorated, and are no longer an aid to scientific progress. Second, if, as Polanyi's work suggests, the standards of scientific rationality are research-immanent, part of the practice of science, rather than an external agency which guides that practice, if their change can be controlled and understood only by those immersed in research, then 'a citizen who wants to judge science must either become a scientist himself, or he must defer to the advice of experts' ([1980], p. 10/*PP2*, p. 26). In such an event, the demand to base political action on scientific standards amounts to *elitism*. This conclusion has been drawn, and accepted, by naturalists like Polanyi and Kuhn. Feyerabend, of course, resists it.

For him, the answer to the question of which standards citizens should use is obvious: '*A citizen will use the standards of the tradition to which (s)he belongs*' (*PP2*, p. 27; see [1980], p. 11). This solution, unlike that of 'the intellectuals' (the friends of rationalism), is democratic. If the rationalist complains that theories from traditions other than science cannot be taken seriously, because they have no results, the reply is that the relevant comparative studies simply have not been carried out:

> The sciences, it is said, are uniformly better than all alternatives – but where is the evidence to support this claim? Where, for example, are the control groups which show the uniform (and not only the occasional) superiority of Western scientific medicine over the medicine of the *Nei Ching*? Or over Hopi medicine? ([1980], p. 13/*PP2*, p. 29)

Furthermore, Feyerabend suggests, the rationalist complaint *presupposes* what are to be established, that is, standards which make the results of science seem worthwhile. By the very different standards of another tradition, the 'achievements' of Western science may seem piffling.

For Feyerabend, the equality embodied in contemporary Western societies is not an equality of traditions, but '*equality of access to one particular tradition* – the tradition of the White Man' ([1980], p. 14/*PP2*, p. 29; *SFS*, p. 9). By contrast, democratic relativism seeks to guarantee equal rights to all traditions. It will of course be objected that this means tolerating certain 'traditions', like Nazism, which liberals and democrats usually oppose, and may even seek to suppress. In response, Feyerabend reassures us that although such traditions must be allowed to exist, this liberty will not carry any licence to impose their form of life on others: every tradition will be protected (by its own members, or by the State's apparatus?) from outside interference. Where exceptions to this rule are necessary, their form and location are decided by specially elected groups of citizens – 'democratic councils' – and not by experts.

Feyerabend then presents arguments in favour of democratic relativism: (a) people have rights, including the right to live as they see fit. No institution should be allowed to force people to accept behaviour they deem unacceptable, or to reject behaviour they value. 'Science or rationalism, in this view, are instruments put at the disposal of the people *to be used by them as they see fit*: they are *not* necessary conditions of rationality, or citizenship, or life. Scientists are salesmen of ideas and gadgets, they are not judges of Truth and Falsehood. Nor are they High Priests of Right Living' ([1980], p. 15/ *PP2*, p. 31). (b) As John Stuart Mill originally indicated, a society containing many traditions side-by-side has a much better chance of judging each tradition than a monistic society. We can learn much from other traditions, including those in 'primitive' societies. Democratic relativism makes the contrasts between their forms of life and ours stand out, and thus allows each group to learn from the other. (c) Scientific views are not only incomplete, they are often *erroneous*: 'Routine arguments and routine procedures are based on assumptions which are inaccessible to the research of the time and often turn out to be either false or nonsensical' ([1980], p. 16/*PP2*, p. 32). Scientists get belligerent when such assumptions are attacked, and defend with their overblown authority ideas which sometimes they cannot even formulate. On some such occasions, laymen (lawyers, for example) have exposed the scientists' incompetence and incomprehension (*AM*[1], p. 307; [1975], p. 162; *SFS*, pp. 88–91). Cases like this suggest that

> *fundamental debates between traditions are debates between laymen which can and should be settled by no higher authority than the authority of laymen, i.e. democratic councils.* ([1980], p. 16/*PP2*, p. 32)

Democratic relativism must be introduced by loosening up the protective machinery of the State, and detaching it from the traditions which today use it exclusively for their own purposes. Feyerabend's ultimate conclusion, therefore, is that just as the Church was separated from the State, so today science should be separated from the State:

> [T]he time is overdue for adding the separation of state and science to the by now quite customary separation of state and church. Science is only *one* of the many instruments man has invented to cope with his surroundings, It is not the only one, it is not infallible, and it has become too powerful, too pushy, and too dangerous to be left on its own. (*AM*[1], pp. 216–17.)[13]

This separation, and the democratic relativism it subserves, will not be imposed 'from above' by a gang of radical intellectuals, but will be realized from within, by those citizens who seek independence, and in the manner of their choosing. Feyerabend sums this up in his slogan: *'Citizens' initiatives instead of philosophy!'* ([1980], p. 17/*PP2*,

p. 33). In the free society he envisages, people first undergo a process of *general* education which prepares them to make a choice between the available standards, but must on no account commit them to the standards of any particular group:

> A mature citizen is not a man who has been *instructed* in a special ideology, such as Puritanism, or critical rationalism, and who now carries this ideology with him like a mental tumour, a mature citizen is a person who has learned how to make up his mind and who has then *decided* in favour of what he thinks suits him best. He is a person with a certain mental toughness . . . and who is therefore able *consciously to choose* the business that seems to be most attractive to him rather than being swallowed by it. (*AM*[1], p. 308. See also p. 218)

This education involves a comparative but purely historical study of major ideologies. Everything taught is of course, for Feyerabend, a theory (or ideology). It is true that some theories can, as he now insists, be taught without being taught as the truth. (This might be thought to betoken a non-realist view of theories.) But Feyerabend does not explain how one could possibly teach children 'theories' like 'Folk Psychology' in this way. He fails to address the question of whether there are 'theories' which are fundamentally and *constitutively* related to human action and our self-conception, things which are both logically and temporally prior to the use of theories (properly so-called), and which can be taught only in a way which prevents the question of their truth being raised.

Just as parents choose the religion that their children are brought up in so, in Feyerabend's free society, parents choose what their children can learn at school (science or magic, the theory of evolution or 'creation science', etc.). Feyerabend incites us to follow the example of the Californian opponents of the theory of evolution, and to free society from the stranglehold of ideologically-petrified science just as our ancestors freed us from the stranglehold of the One True Religion. At the end of such an education, the citizen will be in a position to choose between rationalism and irrationalism, science and myth, science and religion. Feyerabend cannot quite help saying that this choice will be more rational than the uninformed, unprepared, un-conscious and forced 'choice' in favour of science which people usually make. In this respect he can be seen to be arguing, not against rationality *tout court*, but rather in favour of a wider conception of rationality than that embodied in some existing versions of 'rational-ism'.

10.5   The Problem of the Excellence of Science

Feyerabend's political conclusion is reinforced by his arguments against the current status of science, in which he raises what he calls 'the long-forgotten problem of the excellence of science' (AM[1], p. 296). The status quo, that science *is* almost uniformly excellent and therefore deserves our intellectual adherence, as well as our continued financial support, is defended by three arguments. First, it is impressed upon us that modern science is a fact of life. Its competitors, by contrast, have either died out or survive only among isolated pockets of weirdos and primitives. This is supposed to be because science vanquished its opponents at the time of the Scientific Revolution. During that period, science represented the vanguard of enlightenment, pitted against the forces of authoritarianism and superstition.

Feyerabend makes several points in response. Although science was once a force for intellectual liberation, he feels that it is no longer so. He reminds us of the stranglehold of the myth predicament:

> Any ideology that breaks the hold of a comprehensive system of thought contributes to the liberation of man. Any ideology that makes man question inherited beliefs is an aid to enlightenment. A truth that reigns without checks and balances is a tyrant who must be overthrown and any falsehood that can aid us in the overthrow of this tyrant is to be welcomed. It follows that 17th and 18th century science was indeed an instrument of liberation and enlightenment. It does not follow that science is bound to *remain* such an instrument. There is nothing inherent in science or in any other ideology that makes it *essentially* liberating. Ideologies can deteriorate and become stupid religions. Look at Marxism. ([1975], in Hacking [1981, pp. 156–7])

Thankfully, science is not quite omnipresent. Even if it were, this would not add up to an argument for its acceptability: 'if a country is invaded by locusts then it is useful to study their habits, but it would be quite unreasonable to turn them into national deities' ([1980], p. 13/PP2, p. 29). More importantly, although it was indeed a force for enlightenment during the Scientific Revolution, science simply overpowered its opponents, rather than refuting or convincing them. It took over by force, rather than by argument:

> When modern science arose it had some successes and was close to the heart of powerful interest groups. The successes combined with the power gradually eliminated competitors such as alchemy and the magic world view although these competitors had suffered only a temporary setback and although they were still studied by outstanding scientists such as Newton. ... The scientific revolution of the 16th and 17th centuries ... assigned to the temporarily defeated rivals a place outside science and so

prevented their return. Today science is on top because the show has been rigged in its favour and not because of any inherent excellence either of its methods, or of its results. ([1977a], p. 13)

Although science overtook its scholarly rivals in a competition that had some semblance of fairness, it supplanted and removed 'primitive' views like magic and religion mainly through instilling a firm belief in the superiority of the white man and his works. Non-Western cultures have their own science (for example, the astronomy of Stone Age man, folk medicine), and their own share of insights into the nature of things. A fair competition between science and non-scientific views would involve each view having access to an equal share of resources. Such a competition has never been staged, and we cannot even anticipate its outcome.

Feyerabend then argues that the differences between science and myth are superficial, and do not support the provision of a special status for science. Rationalists have not *examined* the comparative merits of science and myth. This has been left to unorthodox anthropologists and historians of ideas. Such examinations, Feyerabend assures us, reveal that

science and myth overlap in many ways, that the differences we think we perceive are often *local* phenomena which may turn into similarities elsewhere and that fundamental discrepancies are results of different *aims* rather than of different methods trying to reach one and the same 'rational' end. (*AM*[1], p. 296)

This can be seen as marking the failure of Feyerabend's project within critical rationalism. Aiming originally to establish the difference between science and myth, Feyerabend finally concludes that there is no important, principled difference between the two: science *is* just the myth of today.

Second, science is defended on the ground that although it is not simply a body of known truths, it is in possession of a special method. The public image of twentieth-century science is determined by technological 'miracles' and by an influential fairy-tale about how they are produced:

According to the fairy-tale the success of science is the result of a subtle, but carefully balanced combination of inventiveness and control. Scientists have *ideas*. And they have special *methods* for improving ideas. The theories of science have passed the test of method. They give a better account of the world than ideas which have not passed the test. The fairy-tale explains why modern society treats science in a special way and why it grants it privileges not enjoyed by other institutions. ... If science has found a method that turns ideologically contaminated ideas into true and useful theories, then it is indeed not mere ideology, but an objective measure of all ideologies. (AM^1, pp. 300–2)

The burden of the central argument of *Against Method* is, of course, that this particular fairy-tale is false. Scientists sometimes solve problems, but not because they follow a scientific methodology. What is more, other people sometimes solve problems too!

Third and lastly, defenders of science sometimes urge that although non-scientific traditions may have 'answers' to Western problems, they cannot be taken seriously because Western institutions alone have results. This is supposed to show that our problems are real, and must be taken seriously by everybody. We already know several Feyerabendian replies to this: that the relevant comparative studies have never been carried out; that the products of science are intrinsically excellent only if one has already adopted standards of scientific rationality as measures of excellence; and that other traditions have produced results too. But Feyerabend also has another line of response, that the results of science have not arisen without any help from non-scientific elements, and are not such that they cannot be improved by the admixture of such elements. Such a defence of science actually relies on the fairy-tale of scientific method, on which Feyerabend has already cast doubt. But he is also keen to give examples of scientific achievements (modern astronomy and dynamics, modern scientific medicine) which could not have advanced without the use of such 'antediluvian' ideas. He takes this rather thin list of inconclusive examples to license the conclusion that:

> *Everywhere* science is enriched by unscientific methods and unscientific results, while procedures which have often been regarded as essential parts of science are quietly suspended or circumvented. (AM^1, p. 305, emphasis added. See also [1975], p. 162)

This bold leap leads Feyerabend not only to advocate the separation of State and science, but also to reaffirm his most radical theoretical and methodological pluralism:

> If we want to understand nature, if we want to master our physical surroundings, then we must use *all* ideas, *all* methods, and not just a small selection of them. (AM^1, p. 306)

Pluralism, the only thing that can be salvaged from the attempt to lay down rules of scientific methodology, Feyerabend now credits to Mill and Boltzmann, rather than to Popper.

10.6   Science and Society

Feyerabend's political philosophy, elaboration of which takes up a large part of his post-*Against Method* output, has suffered a twofold fate. On the one hand, it has been the focus of considerable interest among counter-cultural thinkers of the New Age and the ecology movement (see, for example, some of the essays in Munévar [1991]). On the other hand, it has received little attention from within the academic philosophical community, probably because of a perceived failure to live up to the relevant standards of argument. Philosophical commentators tend to agree that his political theory was not well thought out. Feyerabend, who came to think of both political philosophy and philosophy of science as 'sinks of illiterate self-expression' (*SFS*, p. 10), could hardly have cared less about this. Here I can only indicate some of the most visible questions and problems surrounding his views.[14]

As it stands, democratic relativism says only that in deciding how to evaluate the traditions, institutions and proposals which surround her, the citizen will use the standards of the tradition to which she belongs. Note that it does not say that she *should* use these standards, only that she will do so. This claim is either utterly trivial or false. It depends whether Feyerabend would allow that one *can* make a decision according to a standard of a tradition to which one does not belong. If so, democratic relativism is an empirical conjecture about what would happen in a 'free society', but it is one which we already know will be false. If not, the claim is trivial, and has no ethical implications. What Feyerabend seems to want to say is that people ought to act in accordance with the standards of their own tribe. In not saying this, he does register the obvious problem of self-refutation which lurks just below the surface of many forms of value-relativism.[15] The problem is that value-relativism is a *meta*-ethical view, a view about the nature of moral judgements. It does not entail or even sit comfortably with what we might call 'normative relativism', a complex of ethical views according to which one ought to live by the standards of one's own community, but ought not to interfere with the behaviour or views of members of other communities. There is, in fact, a tension between normative and meta-ethical relativism which Feyerabend fails to resolve. But only the normative ethical view can fund the non-interference demanded by 'democratic relativism'.

Other elements of democratic relativism are equally problematic. The democratic relativist insists on the provision of equal rights for all traditions rather than the provision of equal rights for all individual citizens, because the latter is said to mean in practice 'equal rights of access to positions defined by a special tradition – the tradition of Western Science and Rationalism' (*SFS*, p. 9; [1980], p. 14/*PP2*, p. 29). The State in Feyerabend's Free Society is supposed to be ideologically (and thus epistemologically) neutral. This bears some resemblance to the conception of the liberal State put forward by thinkers in mainstream political philosophy. Despite admiring certain aspects of political anarchism, and certain political and religious anarchists, Feyerabend is definitely not to be numbered among them.[16] Rather, the political philosophy to which he felt most comfortable appealing, and for which he always reserved his most extravagant praise, was the liberalism of John Stuart Mill's *On Liberty*. As well as repeatedly endorsing what he saw as the arguments for pluralism contained therein,[17] Feyerabend at one point favoured a libertarian reading of Mill's text:

> The possibilities of Mill's liberalism can be seen from the fact that it provides room for any human desire, and for any human vice. There are no general principles apart from the principle of minimal interference with the lives of individuals, or groups of individuals who have decided to pursue a common aim. For example, *there is no attempt to make the sanctity of human life a principle that would be binding for all*. Those among us who can realise themselves only by killing their fellow human beings and who feel fully alive only when in mortal danger are permitted to form a subsociety of their own where human targets are selected for the hunt, and are hunted down mercilessly, either by a single individual, or by specially trained groups. So, whoever wants to lead a dangerous life, whoever wants to taste human blood will be permitted to do so within the domain of his own subsociety. *But he will not be permitted to implicate others who are not willing to go his way*. (*SFS*, p. 132/*PP2*, p. 69n).

The basic libertarian idea is that the State, if it must exist, should be a *protective structure* providing only those *essential* public services which would not arise under the operation of the free market. But to libertarians it does not *matter* whether science is excellent, rational and successful, or just ideology! All that matters is that science is not an essential public service which the free market cannot provide. A libertarian can therefore take a short-cut to Feyerabend's conclusion that science should be separated from the State. The argument challenging the excellence of science is, if one adopts this political philosophy, irrelevant.

The other strand in Feyerabend's political philosophy, social democracy, is not well reconciled with his libertarian tendencies. Here, the problems are familiar: does not the cultural autonomy provided by democratic relativism itself not count as an ideology, rather than simply a facilitator of different conceptions of the good? How is it decided which rights are embodied in the protective structure? Can the constitution be changed? What counts as a *single* tradition, and *who* decides what so counts? Is Feyerabend's conception of 'general education' not incompatible with his idea that parents have an unlimited right to choose what their children study? Feyerabend had only the beginnings of answers to some of these questions. His political philosophy will be left to more exacting, but perhaps less exciting, thinkers to work out.

Although Feyerabend became known as a cultural relativist, we have seen that his actual relativist pronouncements are not well thought through. His political stance, 'Democratic relativism', seems to be impotent to avoid the obvious inconsistencies of the most naive form of value-relativism. His defence of Protagorean relativism either leaves Protagoras behind, or falls victim to the same kind of objections Plato levelled against Protagoras' views.

Rather surprisingly, one of the very last things Feyerabend wrote was a general recantation of relativism. Summing up his later feelings about *Against Method* in his autobiography, he admitted that viewing cultures as 'more or less closed entities with their own criteria and procedures' (*KT*, p. 151; see also [1991], p. 508), although consonant with the practice of certain anthropologists, ignored the fact that cultures change and interact. 'Objectivism' *and* relativism, he finally concluded, are *both* 'untenable as philosophies [and] bad guides for a fruitful cultural collaboration' (*KT*, p. 152). He also distanced himself from democratic relativism, on the ground that this view was formed by reflecting on science, common sense, and other cultures *from afar* (see [1991], p. 508).

10.7   Conclusion

Feyerabend was not a careful thinker. His importance lies in the commitment with which, in each of the two major phases of his philosophical work, he sought a general account of human knowledge, drawing arguments and examples from spheres of human activity beyond the imagination of most academic philosophers.

What should we say, finally, about the relationship between science and other intellectual pursuits presented in these different phases? The relationship Feyerabend presented is, I believe, weakened by certain flawed assumptions which persist through both phases. According to him, each kind of activity, whether scientific or otherwise, is premised upon some theory. These theories should always be interpreted in their most uncompromising form, as attempts to describe the underlying nature of (certain parts of) reality. Scientific theories, understood thus, appear to be in flat-out conflict with the theories embedded in non-scientific activities. We are therefore forced to choose between them. To make such a choice, we must refer to standards. If we were to refer only to the kinds of standards embodied in scientific methodology then, perhaps, scientific theories would win out every time. We would therefore have to endorse the super-realist conclusion which, as we saw, Feyerabend did favour during his first phase. Not only our comfortable everyday understanding of ourselves and the world around us, but also the very *concepts* in terms of which this understanding is expressed, would be shattered.

According to the *later* Feyerabend, however, such a verdict would be the product of a biased contest. Rather than choosing on the basis of scientific standards, which are not 'objective' but partial, we ought to choose on the basis of whatever standards we wish. The result will be that some of us continue to favour the conclusions apparently warranted by scientific theories, while others prefer the conclusions licensed by non-scientific theories. There is no question of one of these choices being incorrect, for there is no set of standards common to all the possible standpoints available to compare their merits. And there is no question of one of these choices being more rational than another, since *all* ultimate commitments are non-rational. The resulting situation, however, is *desirable*, since it ensures a desirable plurality of views. We will be set at each others' conceptual and theoretical throats, for ever.

Feyerabend's initial exaggeration of the scope of science, and his later diminishing of it, are two sides of the same coin. I would urge a return to a less glamorous way of seeing the situation, in which science tells us about *some* properties of the familiar objects that surround us, but not about others. This means that some of our pre-scientific views about things are open to correction, while others are not. In particular, science cannot tell us what we mean by the words we use; it can change, but cannot falsify, relations between concepts; and it cannot overturn the very foundations of our conceptual scheme, for the tasks of science only make sense in terms of that scheme. This way of looking at things invalidates Feyerabend's assertion that science has no greater authority than any other form of life. The truth is that while science is a sophisticated instrument for answering certain questions it can, legitimately, contribute nothing to answering others. The real philosophical critique of science is a critique of its pretensions to answer *all* our questions.

Notes

Introduction: Feyerabend's Life and Work

1 The exception is that in ch. 1 I have to discuss some material from later papers in order to identify the foundations of Feyerabend's earlier project.

Chapter 1 Philosophy and the Aim of Science

1 As well as Feyerabend's own 'solution' to the paradox, which implies that a correct analysis does not convey any knowledge which one does not already possess ([1956a], p. 95).
2 'Scientism' being a pejorative term for the attempt to see all cognitive activities on the model of (natural) science.
3 There is something obsessive about the way in which Popper became Feyerabend's favourite whipping-post. In *Science in a Free Society* alone, he characterizes Popper as one whose philosophy of science is a step back from that of Ernst Mach (p. 10), as a 'slavish imitator' of true enlightenment ideas (p. 199n), as a naive rationalist (pp. 26n, 82n, 163), as immodest (p. 59n), as one who turns complex problems into 'trite puzzles' (p. 204), and as an illiterate denigrator of Aristotle (p. 217).
4 That Feyerabend was still very much under the influence of Popper in the mid-1960s is suggested by his gushing and wholly uncritical review of *Conjectures and Refutations*, a book he calls 'a major contribution to philosophy, and [. . .] a major event in the history of this subject' ([1965d], p. 88).
5 See ch. 3 of Popper [1972]. I am not suggesting that there is any commitment on Feyerabend's part to the metaphysical machinery which Popper introduces to explain his project, the three 'worlds'. For Feyerabend's critical attitude to this, see his [1974].
6 Following Newton-Smith [1981], pp. 4ff.
7 For an account and critique of Popper's normative conception of epistemology, see Preston [1994].

8 Kraft [1960], p. 32. The translations used are Feyerabend's own. I have used Feyerabend's account because Kraft thanked Feyerabend for his review's 'careful analysis' of this book (see *KT*, p. 74).

9 Feyerabend refers the reader to precisely those sections of Popper [1959] criticized in my [1994]. Unfortunately, the part of his original footnote which contains this acknowledgement to Popper ([1958a], p. 166, n. 25) is missing from the reprint in *PP*1, p. 34, n. 24. Feyerabend later suppressed many such references to Popper in his early work.

10 Although the original article contains this passage, the version reprinted in *PP*1, pp. 68-9, contains only the first two sentences. This is not the only substantial passage which has been so suppressed, although it is the most important. Passages or comments committing Feyerabend to this normative conception of epistemology occur in (the original versions of) many of his papers prior to about 1968.

11 Of course, philosophers might *suggest* methodological rules for science. But it will only be correct to treat these as rules of scientific method if scientists actually follow them.

12 For further evidence, see [1962a], pp. 38–40, and [1963b]: *PP*1, p. 171. For evidence of Feyerabend's later retreat from the nature/convention dualism see his new footnote to [1960c]: *PP*1, p. 235, n. 20.

13 Feyerabend's attempt to refute the hypothesis will be discussed in ch. 7.

14 Perhaps the best discussion of Feyerabend's ethical basis for philosophy occurs in Hooker [1972].

Chapter 2   Meaning: The Attack on Positivism

1 [1955]: *PP*2, p. 125. Recall that Popper always expressed disdain for the whole 'problem of meaning', complaining that it is 'notorious for its vagueness and hardly offer[s] hope of a solution' ([1963], p. 111).

2 This 'anti-realist' conception was canvassed from the late 1950s onwards by Michael Dummett and others (see Dummett [1978]). For more evidence of Feyerabend's interpretation, see [1966b], p. 8.

3 See Baker & Hacker [1983], ch. 4. In calling this approach 'instrumentalist', I mean only that it identifies the meaning of a word with its role according to a system of rules. I am not suggesting that such instrumentalism embodies a *causal* conception of meaning.

4 See Hacking [1975] and Rorty [1980].

5 See Preston [1995]. For some of the comments which led people to attribute this view to Feyerabend well before he endorsed it, see ch. 6, section 4 this book.

6 See Achinstein [1964] and [1968], Putnam [1965] and Shapere [1966].

7 But where followers of Wittgenstein are berated for producing material that 'more closely resemble[s] the products of pedestrian and second-rate poets rather than those of people who are supposed to *think*, namely, philosophers' ([1960d], p. 247).

8 There is one point where Feyerabend is explicitly sceptical about the contextual theory of meaning. In [1964c] he chides Hanson for his 'extreme contextualism of meaning' (p. 266), deeming it refuted by its implication that independent tests of a theory are logically impossible. But Feyerabend soon

changed his mind about whether this consequence invalidates the contextual theory: see [1965c]: *PP*1, p. 117, n. 34.

9 Newton-Smith [1981], p. 12, for example.

10 Although, as David Papineau points out, this creates problems with the definition of law-likeness ([1979], pp. 42–5).

11 Promising at least in that it does not entail meaning-scepticism. But see Achinstein [1964] and ch. 6 this book for objections and suggestions which ought to return us to an 'instrumentalist' conception.

12 Feyerabend's theory has recently been developed, though, by his followers and others, into 'conceptual-role semantics' and 'state-space semantics'. See e.g. P. M. Churchland [1979] and [1989].

13 Hacking [1983], pp. 41–2.

14 This conception is far from that of the Logical Positivists, for whom neither realism nor positivism can have repercussions for the practice of science. Whether we see our theories in one way or another could make no difference to that practice.

15 The best introduction to Wittgenstein's arguments against a logically private language is Peter Hacker's entry on 'Private Language Argument' in Dancy & Sosa [1992].

Chapter 3 Theories of Observation

1 In lumping Hanson and Hesse together, Feyerabend tends to mix up (3) with (1) and (2).

2 The reference is to Alexander Michotte's experiments on the perception of causality. See also *AM*[1] p. 133, p. 269 n. 126.

3 Ryle [1954], p. 90.

4 For discussion of this issue, see Hacking [1983], pp. 172–6. This should be compared with Mary Hesse's [1974], ch. 1 of which is the most impressive defence of the theory-ladenness thesis.

5 For Feigl's complaints, see Feyerabend [1958a]: *PP*1, pp. 31–2; [1965c]: *PP*1, pp. 116, 121.

6 [1962a]: *PP*1, pp. 49ff; [1965a], p. 152. See also [1965c]: *PP*1, p. 125. By contrast, he rebukes Wittgenstein for *not* having a clear idea of the theory ([1962a], p. 39 n. 27). Feyerabend had a less one-sided view of the Logical Positivists than many of his contemporary 'post-positivist' philosophers. Nevertheless, his attribution of the pragmatic theory of observation to Carnap has been challenged in Oberdan [1990]. The pragmatic theory seems closer to the views of Neurath.

7 See Carnap [1956], section II.

8 This attitude is evinced most clearly by philosophers like Quine, Goodman and Putnam.

9 Feyerabend thinks it 'merely *plain commonsense* (though perhaps not Oxford commonsense) to prefer total improvement to piecemeal tinkering' ([1965c]: *PP*1, p. 127). For the contrast between conservatives and radicals about conceptual change, I am indebted to Jim Duthie.

10 Feyerabend does recognize the existence of grammatical rules, but he persists in treating them as factual claims. Lacking space to develop objections to this, I refer readers to section 4 of Glock [1994].

11 See P. M. Churchland [1979], [1985] and [1989].

12 See his paper 'Meaning', repr. in Grice [1989].
13 [1960a]. This paper developed from Feyerabend's discussions with Herbert
   Feigl (see *KT*, pp. 116–17). When quoting from this paper, of which no English
   translation is generally available, I have used a partial translation by F.
   Buchler and J. Bruck, kindly supplied to me by George Couvalis. But, because
   this translation is incomplete, I give page references to the original German
   version. Couvalis's book ([1989]), and my review of it (Preston [1992b]) also
   treat the issues dealt with here.
14 Berkeley, *Three Dialogues between Hylas and Philonous*, first dialogue.
15 Feyerabend considers this argument stronger than what he calls Wittgenstein's
   'linguistic proof' that sensations are not sense-data. (Presumably he means
   the 'private language argument'.) Where Wittgenstein wants to show that
   there could *be* no such language, Feyerabend allows that there could be, but
   insists that it would not be a suitable observation-language for science.

Chapter 4   Scientific Realism and Instrumentalism

1 Following Newton-Smith [1981], p. 29.
2 Such as the 'convergent epistemological realism' attacked in Laudan [1981] .
3 This implies, problematically, that there could be a testable scientific theory
   and an untestable metaphysical theory which, in describing exactly the same
   state of affairs, have the same descriptive meaning. This runs against
   Feyerabend's ideas about meaning and empirical content. See ch. 5.
4 Sometimes Feyerabend takes instrumentalism to be the *only* available non–
   realist view. But usually he recognizes others, such as that of Carnap [1956].
   To allay the suspicion that instrumentalism about theories follows from the
   instrumentalist view of language mentioned in ch. 2, we need merely remind
   ourselves that *some* of the uses to which language can be put *are* straightforwardly explanatory and descriptive.
5 This must not be confused with the idea that a conventionalist interpretation
   (à la Poincaré) is the correct interpretation of a scientific theory.
6 See e.g., Baker & Hacker [1984], p. 6.
7 The position of van Fraassen [1980] although not strictly instrumentalist, is an
   example.
8 In the original footnote 21 to [1964a] (p. 291). The new footnote to the
   reprinted version in *PP1* (p. 186) ascribes to Popper only an 'attempt' at such
   a refutation.
9 In the original footnotes 19 and 32 to [1964a] (pp. 290, 296). The new footnote
   to the reprinted version in *PP1* (p. 185) ascribes to Popper only an 'analysis
   and criticism' of this idea.
10 As Deweyan compatibilists like Ernest Nagel, Sidney Morgenbesser and
   Richard Rorty would have it.
11 See Musgrave [1991], p. 244.
12 See also [1958b], p. 89.
13 Osiander, for example, makes an explicit plea for theoretical tolerance (see
   Giedymin [1976], p. 187, [1982], p. 99).
14 *Dogmatic* in virtue of believing that 'we can *know for sure* which theories are
   true and which false' (Musgrave [1991], p. 244); *disappointed* in that they
   believed that the Copernican theory was not a true description of reality, and
   thus could at most be an instrument of prediction. Disappointed realists,

however, are not global instrumentalists, they do not, indeed cannot (as we saw earlier), take an instrumentalist view of *all* theories. Giedymin is correct, as against Musgrave, in thinking that Duhem was not an extreme instrumentalist. He may, however, have been wrong in thinking of Duhem as a conventionalist.

15 The Logical Positivists, with their thesis of the 'Unity of Science', are an exception in this respect.

Chapter 5   Theoretical Monism

1 In his most radical anti-naive-empiricist mood ([1969a]), Feyerabend argued that experience need have *no* role in science at all.

2 I discuss Nagel instead of Newton below since Feyerabend regards Nagel's methodology as being, in crucial respects, an updated version of Newton's. For Newton, see Feyerabend [1970d].

3 Although, as we saw in ch. 3, he thought it refuted as a *general* thesis.

4 More than once, Feyerabend actually identifies dogma with absolute truth, and with certainty. See e.g. [1961d], p. 402.

5 I do not deal with Feyerabend's critique of Hempel here, partly because I believe it to be mistaken. See Coffa [1967].

6 His recent sympathizers are more liberal in this respect, having moderated Feyerabend's opposition to reduction and to 'reductionism'.

7 See e.g. Newton-Smith [1981], pp. 129–33. However, Feyerabend thinks that the condition was wielded not just by the logical empiricists, but also by Ernst Mach, Max Born, and by adherents of the Copenhagen Interpretation of quantum mechanics. See [1963a], pp. 11–12.

8 Ch. 2 of Couvalis [1989] is the best exposition of Feyerabend's critique of cumulativism.

9 Worrall [1989b].

10 See P. M. Churchland [1979], § 11, Hooker [1981], and P. S. Churchland [1986], ch. 7.

11 Wrongly, I believe. See Kuhn [1983a] for remarks defending this approach.

12 Kuhn could reply that none of this is to the point, for he was talking about paradigms, not about theories, which Feyerabend insists on discussing. If we amend Feyerabend's argument by substituting 'paradigm' for 'theory', the resulting thesis (that it is always possible for scientists to work on more than one *paradigm* at a time) is harder to establish. But Feyerabend's examples still seem fairly persuasive.

13 Note that Feyerabend idolizes the individual scientific genius, and still downplays all but the critical function of the community aspect of scientific activity, with which Kuhn is preoccupied.

Chapter 6   Incommensurability

1 Where the original article has 'inconsistency', the reprint (*PP1*, pp. 66–7) has 'conflict' only.

2 Major references are: Kuhn [1962] and [1970], Shapere [1966], Giedymin [1970] and [1971], Davidson [1973], Suppe [1977], Doppelt [1978], Musgrave [1978], P. M. Churchland [1979], Devitt [1979], Moberg [1979], Newton-Smith

[1981], Putnam [1981], Brown [1983], Kuhn [1983b], Burian [1984], Collier [1984], Couvalis [1989], Munévar [1991], Hoyningen-Huene [1993], and Sankey [1994].

3 For example, that incommensurability is not susceptible of explicit definition, that it can only be shown and not explicated, etc. See AM^1, p. 225.

4 But Feyerabend and Kuhn were not the first philosophers of science to use the concept. That honour belongs to the 'radical conventionalists' LeRoy and Ajdukiewicz. See Giedymin [1982].

5 See Sellars [1963], Achinstein [1964], [1968], Naess [1964], Putnam [1965], Smart [1965], Scheffler [1966], Shapere [1966], Fine [1967], Giedymin [1970] and Nagel [1979].

6 Ironically, Feyerabend himself had already objected to Carnap's positivistic theory of meaning on the ground that it implied that any change in a theory would constitute a change in the meaning of its theoretical terms. See [1960b]: $PP1$, pp. 39–40.

7 There are suggestive remarks in Paul Churchland's paper in Munévar [1991]. For a Kuhnian attempt to develop a suitable relationship of 'pragmatic incommensurability', see Collier [1984].

8 However, it must be admitted that Feyerabend did eventually move to a position according to which meaning does not matter at all. See Preston [1995].

9 Newton-Smith, for example, for whom 'The thought that theories are incommensurable is the thought that theories simply cannot be compared and consequently there cannot be any rationally justifiable reason for thinking that one theory is better than another' (Newton-Smith [1981], p. 148). And Shapere, who simply equates incommensurability with incomparability ([1966], pp. 37, 55 in Hacking [1981]).

10 See, for example, Butts [1966] and Townsend [1971].

11 Kuhn, too, denied that his incommensurability thesis involved incomparability. See Kuhn [1983b].

12 The Churchlands, who accept that some conceptual systems are incommensurable, are neither relativists nor 'irrationalists'.

13 For other evidence of Feyerabend's anti-relativism see [1960c]: $PP1$, p. 232, where he rejects Bohm's relativism; [1964d], pp. 245–6, where he rejects a relativistic attitude towards Chinese science; and where he ridicules methodological pluralism, and exhibits little sympathy for what he calls 'the snake of relativism' (p. 250). All these rejections are grounded in his commitment to a strongly normative falsificationist methodology. The final section of [1970a], where Feyerabend asks himself whether his views on incommensurability entail subjectivism or relativism, is the first place where he fails to give an unambiguously negative answer.

14 Attempts to compare incommensurable theories on the basis of the concept of reference, sometimes proposed by Feyerabendian sympathizers (Couvalis [1989], Sankey [1994]), ignore the fact that reliance on conceptually unmediated word/world relations (such as reference) is not in keeping with Feyerabend's philosophy.

Chapter 7 Theoretical Pluralism

1 As we saw in ch. 5, Feyerabend originally thought that Popper was not incriminated, since Popper apparently saw the essential role of alternative theories in the process of testing ([1962a], p. 32). But Feyerabend, probably under the influence of Lakatos, changed his mind, and in the reprinted version of this paper (*PP1*, p. 47n) he takes Popper to task for *not* endorsing theoretical pluralism.

2 The argument appears in [1962a], [1963a], [1964a], [1964b], [1965a], [1966a], [1968b], [1968c], [1969b], *AM* and *FTR*.

3 My sketch is heavily indebted to the elegant account in Brush [1968]. But the story of the discovery and explanation of Brownian motion is also described in Nye [1972], Clark [1976], and Maiocchi [1990], and these authors disagree strongly with one another. Maiocchi's paper is the most congenial to Feyerabend's case. For an introduction to the physics of Brownian motion, see Lavenda [1985].

4 See Einstein [1926].

5 Those who think of Feyerabend as a member of the historical school of philosophers of science might like to consider that nowhere in the (more than ten) papers in which he discusses the case of Brownian motion does he give anything like as detailed a sketch as the one I have just given. Considering that this example is the main illustration of the central argument in the first phase of his philosophy, this is somewhat disappointing. It must also be said that Feyerabend's grasp on the history of the case is weak. His analysis came under fire from Laymon [1977], Worrall [1978b] and [1991], and Laudan [1989a]. Their claims that Feyerabend misdescribed either the physics or the history of his main example would have to be considered in any detailed treatment of this analysis. Feyerabend was defended in Couvalis [1988]. A much longer discussion than I can enter into here would be required to do justice to the contributions to this debate.

6 For some reservations about this claim, see Popper [1957], and Nagel [1979], pp. 82–3.

7 The other examples mentioned are: the case of the motion of an electron within the shell of an atom ([1964b]: *PP1*, pp. 205–6); the case of the deviation of the planet Mercury from the orbit as calculated by Newton ([1968b], p. 132); and the idea of 'observer-independence' ([1969b]: *PP1*, p. 158).

8 Feyerabend berates Popperians for interpreting him as suggesting that any two conflicting alternative theories are related in this way, and maintains that his own claim is merely that there exist theories so related (*PP1*, p. 143 n. 13). But his claim that this relation is 'typical' surely has more import than this.

9 In fact, Laudan argues that M does not refute T; but I think this is because he has not spelled out all the relevant conditions.

10 '[E]very statement has the character of a theory, of a hypothesis,' says Popper ([1959], pp. 94–5). See also p. 59 n. *1, and p. 107 n. *3, on the fact that observation statements 'are always *interpretations* of the facts observed; that they are interpretations in the light of theories'. And see p. 111 (1972 Addendum) on basic statements being impregnated with theories.

11 This is among the things he later comes to believe do *not* exist! In [1970a], the methodological principles of tenacity and proliferation, discussed below, are said to be principles 'for mnemonic reasons only' (*PP2*, p. 138). For a vigorous defence of a moderate theoretical pluralism, see Naess [1964].

12 Some of these methodological rules (notably, the principles of proliferation and of tenacity) even seem to survive the introduction of epistemological anarchism. But although it would be unwise to declare that any of them are at this point denied by Feyerabend, his later use of them can usually be interpreted as *ad hominem*.

13 Feyerabend's 'Thesis I' is also supposed to be a formulation of this principle.

14 In [1965c], *PP*1, p. 107, the 'principle of tenacity' is held to be implied by the principle of proliferation.

15 See, for example, [1964d], p. 250, where Feyerabend is sceptical about what he calls the 'New Liberalism' which urges us to let a thousand *methods* bloom.

Chapter 8   Materialism

1 I am grateful to Christopher Hookway for this reference.

2 Although Feyerabend moved away from this reductionist view of sensations, it has recently been rehabilitated in P. M. Churchland [1985].

3 Feyerabend, in drawing attention to a difference between my thought of X and X itself (*PP*1, p. 165), mis-states the issue here.

4 For this conception, see Hacker [1990].

5 Popper was originally excepted as well!

6 See [1963a], p. 27, [1965a], pp. 177–9, *AM*^{1}, pp. 43–4.

7 See [1963b]: *PP*1, pp. 170, 172; [1965a], pp. 150–1, 152.

8 See [1963a], pp. 31–2.

9 See P. M. Churchland [1985], and P. S. Churchland [1986]. However, eliminativism about mental events apparently has a contemporary defender in Daniel Dennett. See Dennett [1981], p. xx.

10 See Wilkes [1984] and, for a reply, Preston [1989].

11 This objection is astutely deployed in Horgan & Woodward [1985].

12 See [1963d], p. 296, where Feyerabend advises materialists to use identity hypotheses to *redefine* our existing psychological vocabulary.

Chapter 9   Science without Method

1 These letters are now part of the Lakatos Collection in the Archive Division of the British Library of Political and Economic Science, at the London School of Economics.

2 Feyerabend considered these replies worth reprinting, under the jaunty title 'Conversations with Illiterates', as part 3 of SFS.

3 Feyerabend disavowed the intention of *establishing* the 'anarchist' conclusion, claiming only that he was trying to 'make it plausible' with the help of examples (*AM*^{3}, p. 1).

4 Explored in Preston [forthcoming b].

5 For this, see especially *AM*^{3}, p. 129.

6 See, for example, Machamer [1973], Finocchiaro [1980], Krige [1980], Chalmers [1986], Andersson [1991] and Pera [1994].

7 See, for example, Zahar [1982], p. 407, and Worrall [1988], [1989a], in which it is argued that science is characterized by a 'fixed core' of methodological rules.

8 It is tempting to combine this view with the claim that we have had to *learn*

about methodology (Laudan [1989b]). But this seems to imply that we have some fixed, core principles of rationality relative to which we can say that we now know more about methodology than we once did (Worrall [1989a]).

9 See, especially, Newton-Smith [1992].

10 For an attempt to develop a position intermediate between 'rationalism' and relativism, see Laudan [1984] and [1989b]. But see Worrall [1988] and [1989a] for resistance.

11 See, for example, the 1980 addendum to [1962a]: *PP1*, p. 47 n. 6.

12 'Norms and demands must be checked by research, not by appeal to theories of rationality' (*SFS*, p. 117). See also ibid., p. 144 n. 7. In a letter to Lakatos, Feyerabend said: 'My actual historical development away from naive falsificationism involves Kuhn, and a brief discussion I had with von Weizsäcker, when I talked in his seminar in Hamburg and when it was demonstrated to me how vicious and useless general rules of procedure might be' (Letter of 7 Aug. 1970).

13 I am thinking of the 'critical modernism' of Dupré [1993], characterized by its anti-reductionism, anti-essentialism and epistemological pluralism.

14 One commentator, having said that he found Feyerabend's approach in the relevant chapter uncongenial and unilluminating, remarked that 'It seems to mark Feyerabend's reconciliation with an earlier philosophical love: Wittgenstein' (Worrall [1978a], p. 290).

15 'The . . . assertion of powerful linguistic influence on world views seems to be generally unfounded. A persuasive idea or an intense need for intellectual endeavour in a particular field is well able to overcome linguistic barriers' (Currie [1970], p. 418). See also Berlin & Kay [1969], and Rosch [1977]. For other philosophical reservations about the linguistic relativity principle, see Black [1962], Walton [1973], and Lukes in Hollis & Lukes (eds) [1982].

16 In a delightful moment during one of his letters to Lakatos, Feyerabend gleefully reveals his new weapon: 'I have found an excellent new gimmick for my chapter on incommensurability: *Greek pots*. Now, try to criticise me in *that* field!' (Letter of Nov. 1972).

17 Feyerabend himself rejects one structural approach in his review of Wolfgang Stegmüller's work (Feyerabend [1977b]), although he gives no reasons for continuing to subscribe to the 'statement view' of theories.

18 See, for example, Martin [1969]. For some of Feyerabend's comments on 'verstehen', see *PP2*, pp. 6, 130.

19 However, Evans-Pritchard insists that the *first* phase of the anthropologist's work is to 'understand the significant overt features of a culture *and to translate them into terms of his own culture*' (Evans-Pritchard [1962], p. 23; emphasis added). This goes exactly contrary to Feyerabend's account of the second phase of the method. Feyerabend also fails to recognize that Evans-Pritchard's conception of anthropological method underwent a substantial change in the 1940s, *between* the texts he refers to. For discussions of Evans-Pritchard's conception, see Evans-Pritchard [1951] and [1962], Jarvie [1964] and Kuper [1983], chs. 3 and 5.

20 To put it bluntly, he seems to have misunderstood the issue. For his remarks on 'radical translation', see *AM*[1], pp. 280, 287. In one of his very last letters to Lakatos, Feyerabend claims to have anticipated Quine's thesis of the indeterminacy of translation in his own paper 'An Attempt at a Realistic Interpretation of Experience' (Letter of 15 Jan. 1974). But this claim is simply evidence that he had failed to understand Quine.

21 Newton-Smith [1981], p. 163.
22 But see Klein [1986], for serious reservations about such considerations.

Chapter 10   Relativism, Rationalism and a Free Society

1 In a parallel passage of AM^1 (p. 45), it is only 'the semblance of absolute truth' that results. See also the original source of this passage in [1965a], p. 179.
2 See, for example, [1970b], p. 329 n. 35; [1975], p. 158.
3 Despite his admiration for Mill, and his own conflation of unfalsifiability with 'absolute truth', Feyerabend recognizes that Mill himself does *not* deny the existence of truth ([1970c], p. 110 n. 49). This aspect of Feyerabend's philosophy derives no support from Mill's *On Liberty*.
4 For later comments on epistemic ideals see *SFS*, p. 125 (where science is held to conflict with idols such as Truth, Honesty, Knowledge, and Reason), *FTR*, p. 63, and *TDK*, p. 53.
5 However, this ought to give us pause for thought. Surely the fixed core could be rational principles which do not yet amount to anything worth calling a distinctively scientific method?
6 For an admirable discussion of this, and an interesting attempt to escape from it, see Bartley [1962], chs. 4–5.
7 An even more explicit use of the dilemma of ultimate commitment occurs in [1984], p. 45: 'When giving reasons, one arrives at either a point of choice or of intuition, i.e. automatic behaviour, the latter once again being a – this time thoughtless – choice.'
8 Whether Wittgenstein's conclusion should be used to buttress irrationalism is another matter. Wittgensteinians sometimes say that we cannot *choose* our basic commitments, but merely accept them unreflectively. Feyerabend takes this to mean that the critical examination of frameworks is impossible. None of this serves as a reply to the irrationalist.
9 Feyerabend expresses admiration for the existentialist Søren Kierkegaard, as for example at *SFS*, p. 163, where he claims Kierkegaard's *Concluding Unscientific Postscript* as an ancestor of his own views on method; or AM^1, p. 175, where he agrees with Kierkegaard that one's scientific activity may weaken one's strength as a human being.
10 Feyerabend's distinction between different versions of rationalism led him further to distinguish 'abstract' from 'historical' traditions (*PP1*, p. 4; *PP2*, pp. 5–10, 19–21), and 'theoretical' from 'empirical' (historical) traditions (*PP2*, pp. 8–12; *FTR*, pp. 118, 166, 294ff). Unfortunately, I have no space to consider this line of development here.
11 For more on Protagorean relativism, see Preston [1992a], [1995], and [forthcoming a].
12 Later in the same book, he disavows the intention of *establishing* relativism in *AM*: 'There is no attempt on my part to show "that an extreme form of relativism is *valid*". . . I merely argue that *the path to relativism has not yet been closed by reason* so that the rationalist cannot object to anyone entering it. Of course, I have considerable *sympathy* for this path and I think it is the path of growth and freedom, but that is a different story' (*SFS*, p. 145). This admission gives some credence to the objection that the relativist's position may be consistent but that it cannot be established since it denies the presupposition

of claims to validity, and is not such that a person can be rationally persuaded to embrace it.

13 On the separation of State and science, see also AM^1 ch. 18; SFS, p. 31 and Part Two.

14 For other problems, see Koertge [1980], Yates [1984], [1985], Alford [1985], Siegel [1989] and Gjertsen [1992].

15 See, for example, Bernard Williams's first paper in Krausz & Meiland [1982].

16 For Feyerabend's objections to political anarchism, see AM^1, pp. 20, 21, 187–9. For his explicit rejection of it, SFS, p. 147 n. 12.

17 See, for example, AM^1, pp. 52–3; $PP2$, pp. 65–73. 'It is not possible', says Feyerabend, 'to improve upon [Mill's] arguments' (SFS, p. 86).

Bibliography

Note: Where works are listed as having been reprinted, page references in the text are to the reprinted version.

Works by Paul Feyerabend

[1955]: 'Wittgenstein's *Philosophical Investigations*', *Philosophical Review*, vol. 64.

[1956a]: 'A Note on the Paradox of Analysis', *Philosophical Studies*, vol. 7.

[1956b]: Review of G. Martin, *Kant's Metaphysics and Theory of Science*, *British Journal for the Philosophy of Science*, vol. 7.

[1957]: 'On the Quantum Theory of Measurement', in S. Körner (ed.), *Observation and Interpretation* (London: Butterworth).

[1958a]: 'An Attempt at a Realistic Interpretation of Experience', *Proceedings of the Aristotelian Society*, vol. 58.

[1958b]: 'Complementarity', *Aristotelian Society*, suppl. vol. 32.

[1958c]: 'Reichenbach's Interpretation of Quantum Mechanics', *Philosophical Studies*, vol. 9.

[1960a]: 'Das Problem der Existenz theoretischer Entitäten', in E. Topitsch (ed.), *Probleme der Wissenschaftstheorie: Festschrift für Viktor Kraft* (Vienna: Springer-Verlag).

[1960b]: 'On the Interpretation of Scientific Theories', in *Proceedings of the 12th International Congress in Philosophy* [1958] (Milan/Venice), vol. 5.

[1960c]: 'Professor Bohm's Philosophy of Nature', *British Journal for the Philosophy of Science*, vol. 10.

[1960d]: '*Patterns of Discovery*, by N. R. Hanson', *Philosophical Review*, vol. 69.

[1961a]: 'Knowledge without Foundations', two lectures delivered on the Nellie Heldt Lecture Fund, Oberlin College, Ohio.

[1961b]: 'Niels Bohr's Interpretation of the Quantum Theory', in H. Feigl & G. Maxwell (eds), *Current Issues in the Philosophy of Science* (New York: Holt, Rinehart & Winston).

[1961c]: Comments on papers by N. R. Hanson, W. Sellars, A. Grünbaum, S. F. Barker and E. L. Hill, in Feigl & Maxwell (ibid.).

[1961d]: 'Metascience' (Review of M. Bunge, *Metascientific Queries* and *Causality*), *Philosophical Review*, vol. 70.

[1962a]: 'Explanation, Reduction, and Empiricism', in H. Feigl & G. Maxwell (eds), *Scientific Explanation, Space, and Time: Minnesota Studies in the Philosophy of Science*, vol. 3 (Minneapolis: University of Minnesota Press).

[1962b]: 'Problems of Microphysics', in R. G. Colodny (ed.), *Frontiers of Science and Philosophy* (Englewood Cliffs, NJ: Prentice-Hall).

[1962c]: 'Problems of Microphysics', reprinted in S. Morgenbesser (ed.), *Philosophy of Science Today* (New York: Basic Books, 1967).

[1963a]: 'How to be a Good Empiricist: A Plea for Tolerance in Matters Epistemological', in B. Baumrin (ed.), *Philosophy of Science: The Delaware Seminar*, vol. 2, (New York: Interscience Press).

[1963b]: 'Materialism and the Mind–Body Problem', *Review of Metaphysics*, vol. 17.

[1963c]: '*Erkenntnislehre*, by Viktor Kraft', *British Journal for the Philosophy of Science*, vol. 13.

[1963d]: 'Comment: "Mental Events and the Brain"', *Journal of Philosophy*, vol. 60.

[1963e]: 'Professor Hartmann's Philosophy of Nature', *Ratio*, vol. 5.

[1964a]: 'Realism ,and Instrumentalism: Comments on the Logic of Factual Support', in M. Bunge (ed.), *The Critical Approach to Science and Philosophy* (New York: Free Press).

[1964b]: 'A Note on the Problem of Induction', *Journal of Philosophy*, vol. 61.

[1964c]: Review of N. R. Hanson, *The Concept of the Positron*, *Philosophical Review*, vol. 73.

[1964d]: Review of A. C. Crombie (ed.), *Scientific Change*, *British Journal for the Philosophy of Science*, vol. 15.

[1965a]: 'Problems of Empiricism', in R. G. Colodny (ed.), *Beyond the Edge of Certainty*, University of Pittsburgh Studies in the Philosophy of Science (Englewood Cliffs, NJ: Prentice-Hall).

[1965b]: 'On the "Meaning" of Scientific Terms', *Journal of Philosophy*, vol. 62.

[1965c]: 'Reply to Criticism: Comments on Smart, Sellars, and Putnam', in R. S. Cohen & M. W. Wartofsky (eds), *Boston Studies in the Philosophy of Science*, vol. 2: *In Honor of Philipp Frank* (New York: Humanities Press).

[1965d]: Review of K. R. Popper, *Conjectures and Refutations*, *Isis*, vol. 56.

[1966a]: 'The Structure of Science' (Review of E. Nagel, *The Structure of Science*), *British Journal for the Philosophy of Science*, vol. 17.

[1966b]: 'Herbert Feigl: A Biographical Sketch', in P. K. Feyerabend & G. Maxwell (eds), *Mind, Matter and Method: Essays in Philosophy and Science in Honor of Herbert Feigl* (Minneapolis: University of Minnesota Press).

[1968a]: 'A Note on Two "Problems" of Induction', *British Journal for the Philosophy of Science*, vol. 19.

[1968b]: 'Science, Freedom, and the Good Life', *Philosophical Forum*, vol. 1(2).

[1968c]: 'Outline of a Pluralistic Theory of Knowledge and Action', in S. Anderson (ed.), *Planning for Diversity and Choice* (Cambridge, Mass.: MIT Press).

[1969a]: 'Science without Experience', *Journal of Philosophy*, vol. 66.

[1969b]: 'Linguistic Arguments and Scientific Method', *Telos*, vol. 2(1).

[1969c]: 'A Note on two "Problems" of Induction', *British Journal for the Philosophy of Science*, vol. 19.

[1970a]: 'Consolations for the Specialist', in I. Lakatos & A. Musgrave (eds), [1970].

[1970b]: 'Problems of Empiricism, Part II', in R. G. Colodny (ed.), *The Nature and Function of Scientific Theory* (Pittsburgh: University of Pittsburgh Press).

[1970c]: 'Against Method: Outline of an Anarchistic Theory of Knowledge', in M. Radner & S. Winokur (eds), *Minnesota Studies in the Philosophy of Science*, vol. 4: *Analyses of Theories and Methods of Physics and Psychology* (Minneapolis: University of Minnesota Press).

[1970d]: 'Classical Empiricism', in R. E. Butts & J. W. Davis (eds), *The Methodological Heritage of Newton* (Oxford: Basil Blackwell).

[1970e]: 'Experts in a Free Society', *The Critic*, vol. 29(2), Nov./Dec.

[1974]: Review of K. R. Popper, *Objective Knowledge*, *Inquiry*, vol. 17.

[*AM*[1]]: *Against Method* (Verso: London, 1975).

[1975]: 'How to Defend Society against Science', *Radical Philosophy*, no. 11. (Reprinted in Hacking (ed.), [1981].)

[1976]: 'On the Critique of Scientific Reason', in C. Howson (ed.), *Method and Appraisal in the Physical Sciences* (Cambridge: Cambridge University Press).

[1977a]: 'Rationalism, Relativism, and Scientific Method', *Philosophy in Context*, suppl. 6(1).

[1977b]: 'Changing Patterns of Reconstruction', *British Journal for the Philosophy of Science*, vol. 28.

[*SFS*]: *Science in a Free Society* (London: New Left Books, 1978).

[1980]: 'Democracy, Elitism, and Scientific Method', *Inquiry*, vol. 23.

[*PP*1]: *Realism, Rationalism, and Scientific Method: Philosophical Papers*, vol. 1 (Cambridge: Cambridge University Press, 1981).

[*PP*2]: *Problems of Empiricism: Philosophical Papers*, vol. 2 (Cambridge: Cambridge University Press, 1981).

[1981]: 'More Clothes from the Emperor's Bargain Basement: A Review of Laudan's *Progress and its Problems*', *British Journal for the Philosophy of Science*, vol. 32.

[1984]: 'Science as Art', *Art & Text*, nos 12 & 13.

[*FTR*]: *Farewell to Reason*, (London: Verso/New Left Books, 1987).

[*AM*[2]] *Against Method*. (2nd ed. London: Verso, 1988).

[*TDK*]: *Three Dialogues on Knowledge* (Oxford: Basil Blackwell, 1991).

[1991]: 'Concluding Unphilosophical Conversation', in Munévar (ed.), [1991].

[*AM*[3]] *Against Method*. (3rd ed. London: Verso, 1993).

[*KT*]: *Killing Time: The Autobiography of Paul Feyerabend* (Chicago: University of Chicago Press, 1995).

[1995a]: 'Two Letters of Paul Feyerabend to Thomas S. Kuhn on a Draft of *The Structure of Scientific Revolutions*', *Studies in History and Philosophy of Science*, vol. 26(3).

Other Works Cited

Achinstein, P. [1964]: 'On the Meaning of Scientific Terms', *Journal of Philosophy*, vol. 61.

—— [1968]: *Concepts of Science* (Baltimore: Johns Hopkins University Press).

Agassi, J. [1976]: Review of *Against Method*, *Philosophia*, vol. 6.

Alford, C. F. [1985]: 'Yates on Feyerabend's Democratic Relativism', *Inquiry*, vol. 28.

Andersson, G. [1991]: 'The Tower Experiment and the Copernican Revolution', *International Studies in the Philosophy of Science*, vol. 5.

Austin, J. L. [1979]: 'A Plea for Excuses', in his *Philosophical Papers*, 3rd edn (Oxford: Oxford University Press).

Baker, G. P. & Hacker, P. M. S. [1983]: *Wittgenstein: Meaning and Understanding*. (Oxford: Basil Blackwell).

—— [1984]: *Scepticism, Rules and Language*. (Oxford: Basil Blackwell).

Bartley, W. W. III, [1962]: *The Retreat to Commitment* (New York: Alfred A. Knopf).

Berlin, B. & Kay, P. [1969]: *Basic Color Terms* (Berkeley: University of California Press).

Black, M. [1962]: 'Linguistic Relativity: The Views of Benjamin Lee Whorf', in his *Models and Metaphors* (Ithaca, NY: Cornell University Press).

Brown, H. I. [1983]: 'Incommensurability', *Inquiry*, vol. 26.

Brush, S. G. [1968]: 'A History of Random Processes I: Brownian Movement from Brown to Perrin', *Archive for the History of the Exact Sciences*, vol. 5.

Burian, R. M. [1984]: 'Scientific Realism and Incommensurability: Some Criticisms of Kuhn and Feyerabend', in R. S. Cohen & M. W. Wartofsky (eds), *Methodology, Metaphysics and the History of Science* (Dordrecht: D. Reidel).

Butts, R. E. [1966]: 'Feyerabend and the Pragmatic Theory of Observation', *Philosophy of Science*, vol. 33.

Carnap, R. [1956]: 'The Methodological Character of Theoretical Concepts', in H. Feigl & M. Scriven (eds), *Minnesota Studies in the Philosophy of Science*, vol. 1: *The Foundations of Science and the Concepts of Psychology and Psychoanalysis*. (Minneapolis: University of Minnesota Press).

Chalmers, A. [1986]: 'The Galileo that Feyerabend Missed: An Improved Case against Method', in J. A. Schuster & R. R. Yeo (eds), *The Politics and Rhetoric of Scientific Method* (Dordrecht: D. Reidel).

Churchland, P. M. [1979]: *Scientific Realism and the Plasticity of Mind* (Cambridge: Cambridge University Press).

—— [1985]: 'Reduction, Qualia, and the Direct Introspection of Brain States', *Journal of Philosophy*, vol. 82.

—— [1989]: *A Neurocomputational Perspective: The Nature of Mind and the Structure of Science* (Cambridge, Mass.: MIT Press).

Churchland, P. S. [1986]: *Neurophilosophy: Toward a Unified Science of the Mind/ Brain*. (Cambridge Mass.: MIT Press).

Clark, P. [1976]: 'Atomism Versus Thermodynamics', in C. Howson (ed.), *Method and Appraisal in the Physical Sciences* (Cambridge: Cambridge University Press).

Coffa, J. A. [1967]: 'Feyerabend on Explanation and Reduction', *Journal of Philosophy*, vol. 64.

Collier, J. [1984]: 'Pragmatic Incommensurability', in P. D. Asquith & P. Kitcher (eds), *PSA 1984*, vol 1 (East Lansing, Mich.: Philosophy of Science Association).

Couvalis, S. G. [1988]: 'Feyerabend and Laymon on Brownian Motion', *Philosophy of Science*, vol. 55.

—— [1989]: *Feyerabend's Critique of Foundationalism* (Aldershot: Avebury Press).

Crombie, A. C. (ed.) [1961]: *Scientific Change* (London: Heinemann).

Currie, I. D. [1970]: 'The Sapir–Whorf Hypothesis', in J. E. Curtis & J. W. Petras (eds), *The Sociology of Knowledge: A Reader* (London: Duckworth).

Dancy, J. & Sosa, E. (eds), [1992]: *A Companion to Epistemology* (Oxford: Basil Blackwell).

Davidson, D. [1973]: 'On the Very Idea of a Conceptual Scheme', *Proceedings of the American Philosophical Association*, vol. 47. (Reprinted in Krausz & Meiland [1982].)

Dennett, D. C. [1981]: *Brainstorms: Philosophical Essays on Mind and Psychology*. (Brighton: Harvester Press).

Devitt, M. [1979]: 'Against Incommensurability', *Australasian Journal of Philosophy*, vol. 57.

Doppelt, G. [1978]: 'Kuhn's Epistemological Relativism: An Interpretation and Defence', *Inquiry*, vol. 21.

Dummett, M. A. E. [1978]: *Truth and Other Enigmas* (London: Duckworth).

Dupré, J. [1993]: *The Disorder of Things: Metaphysical Foundations of the Disunity of Science* (Cambridge, Mass.: Harvard University Press).

Eddington, A. S. [1928]: *The Nature of the Physical World* (Cambridge: Cambridge University Press).

Einstein, A. [1926]: *Investigations on the Theory of the Brownian Movement* (ed. R. Fürth). (London: Methuen; repr. New York: Dover Publications, 1956).

Evans-Pritchard, E. E. [1951]: *Social Anthropology* (London: Cohen & West).

—— [1962]: *Essays in Social Anthropology* (London: Faber & Faber).

Fine, A. [1967]: 'Consistency, Derivability, and Scientific Change', *Journal of Philosophy*, vol. 64.

Finocchiaro, M. [1980]: *Galileo and the Art of Reasoning: Rhetorical Foundations of Logic and Scientific Method* (Dordrecht: Kluwer, 1980).

Giedymin, J. [1970]: 'The Paradox of Meaning Variance', *British Journal for the Philosophy of Science*, vol. 21.

—— [1971]: 'Consolations for the Irrationalist?', *British Journal for the Philosophy of Science*, vol. 22.

—— [1976]: 'Instrumentalism and its Critique: A Reappraisal', in R. S. Cohen, P. K. Feyerabend & M. Wartofsky (eds), *Essays in Memory of Imre Lakatos* (Dordrecht: D. Reidel).

—— [1982]: *Science and Convention: Essays on Henri Poincaré's Philosophy of Science and the Conventionalist Tradition* (Oxford: Pergamon Press).

Gjertsen, D. [1992]: *Science and Philosophy, Past and Present* (London: Penguin).

Glock, H.-J. [1994]: 'Wittgenstein vs. Quine on Logical Necessity', in S. Teghrarian (ed.), *Wittgenstein and Contemporary Philosophy* (Bristol: Thoemmes).

Grice, H. P. [1989]: *Studies in the Way of Words* (Cambridge, Mass.: Harvard University Press).

Hacker, P. M. S. [1990]: *Wittgenstein: Meaning and Mind* (Oxford: Basil Blackwell).

Hacking, I. [1975]: *Why Does Language Matter to Philosophy?* (Cambridge: Cambridge University Press).

—— (ed.) [1981]: *Scientific Revolutions* (Oxford: Oxford University Press).

—— [1983]: *Representing and Intervening* (Cambridge: Cambridge University Press).

Hanson, N. R. [1958]: *Patterns of Discovery* (Cambridge: Cambridge University Press).

Hesse, M. B. [1974]: *The Structure of Scientific Inference* (London: Macmillan).

—— [1980]: *Revolutions and Reconstructions in the Philosophy of Science* (Brighton: Harvester Press).

Hollis, M. & Lukes, S. (eds) [1982]: *Rationality and Relativism* (Oxford: Basil Blackwell).

Hooker, C. [1972]: Critical Notice of M. Radner & S. Winokur, *Analyses of Theories and Methods of Physics and Psychology*, Minnesota Studies in the Philosophy of Science, vol. 4, *Canadian Journal of Philosophy*, vol. 1.

—— [1981]: 'Toward a General Theory of Reduction' [Parts 1–3], *Dialogue*, vol. 20.

Horgan, T. & Woodward, J. [1985]: 'Folk Psychology is Here to Stay', *Philosophical*

Review, vol. 94.

Hoyningen-Huene, P. [1993]: *Reconstructing Scientific Revolutions: Thomas S. Kuhn's Philosophy of Science* (Chicago/London: University of Chicago Press).

Jarvie, I. C. [1964]: *The Revolution in Anthropology* (London: Routledge).

Klein, W. [1986]: *Second Language Acquisition* (Cambridge: Cambridge University Press).

Koertge, N. [1980]: Review of *Science in a Free Society*, *British Journal for the Philosophy of Science*, vol. 31.

Kraft, V. [1960]: *Erkenntnislehre* (Vienna: Springer-Verlag).

Krausz, M. & Meiland, J. W. (eds) [1982]: *Relativism: Cognitive and Moral* (Notre Dame, Ind.: University of Notre Dame Press).

Krige, J. [1980]: *Science, Revolution and Discontinuity* (Brighton: Harvester Press).

Kuhn, T. S. [1961]: 'The Function of Dogma in Scientific Research', in Crombie [1961].

—— [1962]: *The Structure of Scientific Revolutions* (Chicago: University of Chicago Press).

—— [1970]: *The Structure of Scientific Revolutions*, 2nd edn (Chicago: University of Chicago Press).

—— [1983a]: 'Rationality and Theory Choice', *Journal of Philosophy*, vol. 80.

—— [1983b]: 'Commensurability, Comparability, Communicability', in P. D. Asquith & T. Nickles (eds), *PSA 1982*, vol. 2 (East Lansing, Mich.: Philosophy of Science Association).

Kuper, A. [1983]: *Anthropology and Anthropologists: The Modern British School*, rev. edn (London: Routledge).

Lakatos, I. & Musgrave, A. (eds) [1970]: *Criticism and the Growth of Knowledge* (Cambridge: Cambridge University Press).

Laudan, L. [1977]: *Progress and its Problems: Towards a Theory of Scientific Growth* (Berkeley: University of California Press).

—— [1981]: 'A Confutation of Convergent Realism', *Philosophy of Science*, vol. 48.

—— [1984]: *Science and Values: The Aims of Science and their Role in Scientific Debate* (Berkeley: University of California Press).

—— [1989a]: 'For Method: Or, Against Feyerabend', in J. R. Brown & J. Mittelstrass (eds), *An Intimate Relation* (Dordrecht: Kluwer).

—— [1989b]: 'If it Ain't Broke, Don't Fix it', *British Journal for the Philosophy of Science*, vol. 40.

Lavenda, B. [1985]: 'Brownian Motion', *Scientific American*, no. 252.

Laymon, R. [1977]: 'Feyerabend, Brownian Motion, and the Hiddenness of Refuting Facts', *Philosophy of Science*, vol. 44.

Machamer, P. [1973]: 'Feyerabend and Galileo: The Interaction of Theories and the Reinterpretation of Experience', *Studies in History and Philosophy of Science*, vol. 4.

Maiocchi, R. [1990]: 'The Case of Brownian Motion', *British Journal for the History of Science*, vol. 23.

Martin, J. [1969]: 'Another Look at the Doctrine of Verstehen', *British Journal for the Philosophy of Science*, vol. 20.

Moberg, D. W. [1979]: 'Are there Rival, Incommensurable Theories?', *Philosophy of Science*, vol. 46.

Munévar, G. (ed.) [1991]: *Beyond Reason: Essays on the Philosophy of Paul Feyerabend* (Dordrecht: Kluwer).

Musgrave, A. [1978]: 'How to Avoid Incommensurability', in I. Niiniluoto & R. Tuomela (eds), *The Logic and Epistemology of Scientific Change*, Acta Philosophica

Fennica, 30 (Amsterdam: North-Holland).
—— [1991]: 'The Myth of Astronomical Instrumentalism', in Munévar (ed.) [1991].
Naess, A. [1964]: 'Pluralistic Theorizing in Physics and Philosophy', *Danish Yearbook of Philosophy*, vol. 1.
Nagel, E. [1949]: 'The Meaning of Reduction in the Natural Sciences', repr. in A. Danto & S. Morgenbesser (eds), *Philosophy of Science* (New York: World Publishing).
Nagel, E. [1961]: *The Structure of Science* (London: Routledge & Kegan Paul).
—— [1979]: *Teleology Revisited, and Other Essays in the Philosophy and History of Science* (New York: Columbia University Press, 1979).
Newton-Smith, W. H. [1981]: *The Rationality of Science* (London: Routledge & Kegan Paul).
—— [1992]: 'The Rationality of Science: Why Bother?', in W. H. Newton-Smith & J. Tianji (eds), *Popper in China* (London: Routledge).
Nye, M. J. [1972]: *Molecular Reality: A Perspective on the Scientific Work of Jean Perrin* (London: Macdonald).
Oberdan, T. [1990]: 'Positivism and the Pragmatic Theory of Observation', in A. Fine, M. Forbes & L. Wessels (eds), *PSA 1990*, vol. 1 (East Lansing, Mich.: Philosophy of Science Association).
Papineau, D. [1979]: *Theory and Meaning* (Oxford: Oxford University Press).
Pera, M. [1994]: *The Discourses of Science* (Chicago: University of Chicago Press).
Poincaré, H. [1905]: *Science and Hypothesis*. (Repr. New York: Dover Publications, 1952.)
Popper, K. R. [1945]: *The Open Society and its Enemies*, vol. 1: *The Spell of Plato* (London: Routledge & Kegan Paul).
—— [1957]: 'Irreversibility; or, Entropy since 1905', *British Journal for the Philosophy of Science*, vol. 8.
—— [1959]: *The Logic of Scientific Discovery* (London: Hutchinson).
—— [1963]: *Conjectures and Refutations: The Growth of Scientific Knowledge* (London: Routledge & Kegan Paul).
—— [1972]: *Objective Knowledge: An Evolutionary Approach* (Oxford: Oxford University Press).
Preston, J. M. [1989]: 'Folk Psychology as Theory or Practice? The Case for Eliminative Materialism', *Inquiry*, vol. 32.
—— [1992a]: 'On Some Objections to Relativism', *Ratio*, vol. 5.
—— [1992b]: Review of Couvalis [1989], *International Studies in the Philosophy of Science*, vol. 6.
—— [1994]: 'Methodology, Epistemology and Conventions: Popper's Bad Start', in D. Hull, M. Forbes & R. M. Burian (eds), *PSA 1994*, vol. 1 (East Lansing, Mich.: Philosophy of Science Association).
—— [1995]: 'Frictionless Philosophy: Paul Feyerabend and Relativism', *History of European Ideas*, vol. 20.
—— [forthcoming a]: 'Feyerabend's Final Relativism', *The European Legacy: Toward New Paradigms*, vol. 1.
—— [forthcoming b]: 'Feyerabend's Polanyian Turn'.
Putnam, H. [1965]: 'How Not to Talk about Meaning', in R. Cohen & M. W. Wartofsky (eds) *Boston Studies in the Philosophy of Science*, vol. 2: *In Honor of Philipp Frank* (New York: Humanities Press). (Repr. in Putnam [1975].)
—— [1975]: *Mind, Language, and Reality: Philosophical Papers*, vol. 2 (Cambridge: Cambridge University Press).

—— [1981]: *Reason, Truth, and History* (Cambridge: Cambridge University Press).

Rorty, R. [1980]: *Philosophy and the Mirror of Nature* (Oxford: Basil Blackwell).

Rosch, E. [1977]: 'Linguistic Relativity', in P. N. Johnson-Laird & P. C. Wason (eds), *Thinking* (Cambridge: Cambridge University Press).

Ryle, G. [1954]: *Dilemmas* (Cambridge: Cambridge University Press).

Sankey, H. [1994]: *The Incommensurability Thesis*. (Aldershot: Avebury Press).

Schaffner, K. [1967]: 'Approaches to Reduction', *Philosophy of Science*, vol. 34.

Scheffler, I. [1966]: *Science and Subjectivity*. (Indianapolis: Hackett).

Schilpp, P. A. (ed.) [1949]: *Albert Einstein: Philosopher-Scientist*, (La Salle, Ill.: Open Court).

Sellars, W. [1963]: *Science, Perception and Reality* (London: Routledge & Kegan Paul).

Shapere, D. [1966]: 'Meaning and Scientific Change', repr. in Hacking [1981].

Short, T. L. [1980]: 'Peirce and the Incommensurability of Theories', *Monist*, vol. 63.

Siegel, H. [1989]: 'Farewell to Feyerabend', *Inquiry*, vol. 32.

Smart, J. J. C. [1965]: 'Conflicting Views about Explanation', in R. S. Cohen & M. Wartofsky (eds), *Boston Studies in the Philosophy of Science, Vol. 2: In Honor of Philipp Frank* (New York: Humanities Press).

Strawson, P. F. [1959]: *Individuals: An Essay in Descriptive Metaphysics* (London: Methuen).

Suppe, F. (ed.) [1977]: *The Structure of Scientific Theories* (Urbana: University of Illinois Press).

Townsend, B. [1971]: 'Feyerabend's Pragmatic Theory of Observation and the Comparability of Alternative Theories', in R. C. Buck & R. S. Cohen (eds), *Boston Studies in the Philosophy of Science*, vol. 8: *PSA 1970* (Dordrecht: D. Reidel).

van Fraassen, B. C. [1980]: *The Scientific Image* (Oxford: Oxford University Press).

Walton, K. [1973]: 'Linguistic Relativity', in G. Pearce & P. Maynard (eds), *Conceptual Change* (Dordrecht: D. Reidel).

Whorf, B. L. [1956]: *Language, Thought, and Reality* (Cambridge, Mass.: MIT Press).

Wilkes, K. V. [1984]: 'Pragmatics in Science and Theory in Common Sense', *Inquiry*, vol. 27.

Wittgenstein, L. [1953]: *Philosophical Investigations* (Oxford: Basil Blackwell).

Worrall, J. [1978a]: 'Against Too Much Method', *Erkenntnis*, vol. 13.

—— [1978b]: 'Is the Empirical Content of a Theory Dependent on its Rivals?' in I. Niiniluoto and R. Tuomela (eds), *The Logic and Epistemology of Scientific Change*, Acta Philosophica Fennica, 30 (Amsterdam: North-Holland), pp. 2–4.

—— [1988]: 'The Value of a Fixed Methodology', *British Journal for the Philosophy of Science*, vol. 39.

—— [1989a]: 'Fix it and Be Damned: A Reply to Laudan', *British Journal for the Philosophy of Science*, vol. 40.

—— [1989b]: 'Structural Realism: The Best of Both Worlds?', *Dialectica*, vol. 43.

—— [1991]: 'Feyerabend and the Facts', in Munévar (ed.) [1991].

Yates, S. [1984]: 'Feyerabend's Democratic Relativism', *Inquiry*, vol. 27.

—— [1985]: 'More on Democratic Relativism: A Response to Alford', *Inquiry*, vol. 28.

Zahar, E. [1982]: 'Feyerabend on Observation and Empirical Content', *British Journal for the Philosophy of Science*, vol. 33.

Index